Advances in Intelligent Systems and Computing

Volume 219

T0137836

Series Editor

J. Kacprzyk, Warsaw, Poland

For further volumes:
http://www.springer.com/series/11156

Ad van Berlo · Kasper Hallenborg
Juan M. Corchado Rodríguez · Dante I. Tapia
Paulo Novais
Editors

Ambient Intelligence – Software and Applications

4th International Symposium
on Ambient Intelligence (ISAmI 2013)

 Springer

Editors

Ad van Berlo
Smart Homes - Dutch Expert Centre for Smart
    Living and E-health
LV Eindhoven
The Netherlands

Kasper Hallenborg
The Maersk Mc-Kinney Moller Institute
University of Southern Denmark
Odense M
Denmark

Juan M. Corchado Rodríguez
Departamento de Informática y Automática
Facultad de Ciencias
Universidad de Salamanca
Salamanca
Spain

Dante I. Tapia
Departamento de Informática y Automática
Facultad de Ciencias
Universidad de Salamanca
Salamanca
Spain

Paulo Novais
Departamento de Informática
Universidade do Minho
Campus de Gualtar
Braga
Portugal

ISSN 2194-5357   ISSN 2194-5365   (electronic)
ISBN 978-3-319-00565-2   ISBN 978-3-319-00566-9   (eBook)
DOI 10.1007/978-3-319-00566-9
Springer Cham Heidelberg New York Dordrecht London

Library of Congress Control Number: 2013937323

Printed on acid-free paper

Springer is part of Springer Science+Business Media (www.springer.com)

Preface

This volume contains the proceedings of the 4th International Symposium on Ambient Intelligence (ISAmI 2013). The Symposium was held in Salamanca, Spain on May 22nd-24th at the University of Salamanca, under the auspices of the Bioinformatic, Intelligent System and Educational Technology Research Group (http://bisite.usal.es/) of the University of Salamanca.

ISAmI has been running annually and aiming to bring together researchers from various disciplines that constitute the scientific field of Ambient Intelligence to present and discuss the latest results, new ideas, projects and lessons learned, namely in terms of software and applications, and aims to bring together researchers from various disciplines that are interested in all aspects of this area.

Ambient Intelligence is a recent paradigm emerging from Artificial Intelligence, where computers are used as proactive tools assisting people with their day-to-day activities, making everyone's life more comfortable.

ISAmI 2013 received a total submission of 43 papers from 19 different countries. Each paper has been reviewed by, at least, three different reviewers, from an international committee composed of 50 members from 20 countries. After a careful review, 30 papers were selected to be presented at the conference and published in the proceedings.

Acknowledgments

A special thanks to Alberto Prado, vice-President of Digital Innovation at Philips that was our invited speaker with the talk "Ambient Intelligence innovation: Overview of Philips activity and outlook".

We want to thank all the sponsors of ISAmI'13: IEEE Sección España, CNRS, AFIA, AEPIA, APPIA, and Junta de Castilla y León.

ISAmI would not have been possible without an active Program Committee. We would like to thanks all the members for their time and useful comments and recommendations.

We would like also to thank all the contributing authors and the Local Organizing Committee for their hard and highly valuable work.

Your work was essential to the success of ISAmI 2013.

May 2013

The editors
Ad van Berlo
Kasper Hallenborg
Juan M. Corchado Rodríguez
Dante I. Tapia
Paulo Novais

Organization

Scientific Committee Chairs

Ad van Berlo   Smart Homes (The Netherlands)
Kasper Hallenborg   Univ. of Southern Denmark (Denmark)

Organizing Committee Chair

Juan M. Corchado   University of Salamanca (Spain)
Paulo Novais   University of Minho (Portugal)
Dante I. Tapia   University of Salamanca (Spain)

Program Committee

Ad van Berlo (Co-chairman)   Smart Homes (The Netherlands)
Kasper Hallenborg
  (Co-chairman)   Univ. of Southern Denmark (Denmark)
Andreas Riener   Johannes Kepler University Linz (Austria)
Antonio Fernández Caballero   University of Castilla-La Mancha (Spain)
Benjamín Fonseca   Universidade de Trás-os-Montes e Alto Douro
                                      (Portugal)
Carlos Bento   University of Coimbra (Portugal)
Cecilio Angulo   Polytechnic University of Catalonia (Spain)
Cesar Analide   University of Minho (Portugal)
Davy Preuveneers   Katholieke Universiteit Leuven (Belgium)
Emilio S. Corchado   University of Burgos (Spain)
Flavia Delicato   Universidade Federal do Rio de Janeiro (Brasil)
Florentino Fdez-Riverola   University of Vigo (Spain)
Francesco Potortì   ISTI-CNR institute (Italy)
Francisco Silva   Maranhão Federal University (Brazil)
Fulvio Corno   Politecnico di Torino (Italy)

Goreti Marreiros	Polytechnic of Porto (Portugal)
Gregor Broll	DOCOMO Euro-Labs (Germany)
Guillaume Lopez	University of Tokyo (Japan)
Habib Fardoum	University of Castilla-La Mancha (Spain)
Hans W. Guesgen	Massey University (New Zealand)
Ichiro Satoh	National Institute of Informatics Tokyo (Japan)
Jaderick Pabico	University of the Philippines Los Baños (Philippines)
Javier Carbo	Univ. Carlos III of Madrid (Spain)
Javier Jaen	Polytechnic University of Valencia (Spain)
Joel Rodrigues	University of Beira Interior (Portugal)
José M. Molina	University Carlos III of Madrid (Spain)
José Machado	University of Minho (Portugal)
Juan A. Botía	University of Murcia (Spain)
Junzhong Gu	East China Normal University (China)
Kristof Van Laerhoven	TU Darmstadt (Germany)
Latif Ladid	University of Luxembourg (Luxembourg)
Lourdes Borrajo	University of Vigo (Spain)
Martijn Vastenburg	Delft University of Technology (The Netherlands)
Nuno Garcia	Lusophone University (Portugal)
Óscar García	University of Salamanca (Spain)
Paul Lukowicz	University of Passau (Germany)
Paulo Novais	University of Minho (Portugal)
Radu-Daniel Vatavu	University "Stefan cel Mare" of Suceava (Romania)
Rene Meier	Lucerne University of Applied Sciences (Switzerland)
Ricardo Costa	Polytechnic of Porto (Portugal)
Ricardo S. Alonso	University of Salamanca (Spain)
Rui José	University of Minho (Portugal)
Simon Egerton	Monash University (Malaysia)
Teresa Romão	New University of Lisbon (Portugal)
Tibor Bosse Vrije	Universiteit Amsterdam (The Netherlands)
Veikko Ikonen	VTT Technical Research Centre (Finland)
Vicente Julián	Valencia University of Technology (Spain)
Yi Fang	Purdue University (USA)
Thomas Hermann	Bielefeld University (Germany)
Lawrence Wong Wai Choong	National University of Singapore (Singapore)

Local Organization Committee

Juan M. Corchado	University of Salamanca (Spain)
Paulo Novais	University of Minho (Portugal)
Javier Bajo	Pontifical University of Salamanca (Spain)
Dante I. Tapia	University of Salamanca (Spain)
Fernando de la Prieta	University of Salamanca (Spain)
Cesar Analide	University of Minho (Portugal)

Contents

Content-Based Design and Implementation of Ambient Intelligence Applications

Jurriaan van Diggelen[1], Marc Grootjen[2], Emiel M. Ubink[1],
Maarten van Zomeren[3], and Nanja J.J.M. Smets[1]

[1] TNO, The Netherlands
[2] EagleScience, The Netherlands
[3] Curiehom, The Netherlands
{jurriaan.vandiggelen,emiel.ubink,nanja.smets}@tno.nl,
m.grootjen@eaglescience.nl, maarten@curiehom.nl

Abstract. Optimal support of professionals in complex ambient task environments requires a system that delivers the Right Message at the Right Moment in the Right Modality: $(RM)^3$. This paper describes a content-based design methodology and an agent-based architecture to enable real time decisions of information presentation according to $(RM)^3$. We will describe the full development cycle of design, development, and evaluation. Ontologies are regarded as key enablers as they define the classes and attributes to allow $(RM)^3$ delivery. As a case study, we describe an ambient computing application for human-robot interaction in the Urban Search and Rescue domain.

1 Introduction

Modern Ambient Intelligence (AmI) applications that support professionals in their complex task environments are characterized by a large number of humans, networked computing devices, sensors, and possibly robots working together. All these actors can be sources of valuable information to the professionals that work in this environment, but forwarding all information to everybody will cause unmanageable overload and interruptions. To make optimal use of the vast amounts of digitally stored information we need a system that delivers the Right Message at the Right Moment in the Right Modality, or $(RM)^3$ [1] .

Whereas our aim might appear straightforward, the inherent design complexity is overwhelming. To decide what qualifies as *right*, the system must take into consideration the type of *content* (e.g. whether it is an urgent message or not), the *user* (e.g. current activity, cognitive task load, location, role in the organization, and other context factors [8]), and the *technological environment* (e.g. which interaction devices are currently available to whom, and in which way). All of these factors are highly dynamic and must be decided at runtime. Current design methodologies fall short in adequately addressing these challenges. Software engineering methodologies focus too much on technology and its functional requirements to take usage issues (such as information overload) into account [2]. User-centered design methodologies solve a part of the problem by adopting a view which is centered around humans and

A. van Berlo et al. (Eds.): *Ambient Intelligence – Software & Applications*, AISC 219, pp. 1–8.
DOI: 10.1007/978-3-319-00566-9_1   © Springer International Publishing Switzerland 2013

the tasks they perform (e.g. [3,4]). However such methods do not lead to development of implementable rules or algorithms for distributing messages in an (RM)^3 way.

To bridge the gap between high-level task analysis and an implementable system specification, we have extended the user-centered design methodology called situated Cognitive Engineering (sCE) [5,6] with a content-based perspective and developed an agent-based software architecture which can be used to distribute content according to (RM)^3 in various applications. The purpose of this paper is to present the design methodology, the software architecture, and to demonstrate its functioning in a working prototype.

Our design methodology adopts *content* as the departure point. We regard pieces of content as competing with each other for access to the user interfaces that are available and ultimately, to catch the user's attention. The user's cognitive abilities as well as the various user interfaces are considered as resources that need to be distributed between the tasks that the user is performing and all the information streams that can be consumed by the user. If, how and when information is delivered to the user is made dependent on the user's state, task, context, and available devices and services.

We propose an agent-based architecture that makes these decisions at runtime. A central component in this architecture is a set of ontologies which describe the classes and relationships which form the vocabulary for describing content, users and devices. From a content-based perspective, these ontologies specify the blueprint of the whole application. We apply the framework to a robot case in the domain of Urban Search and Rescue (USAR).

The paper is organized as follows. Section 2 describes the content-based design methodology. Section 3 describes the software architecture for designing content-based AmI applications, which is based on the ontologies constructed in the design phase. Section 4 describes a demonstrator in the USAR domain. We conclude in Section 5.

2   Developing the Design Specification

Following the sCE methodology, we iteratively specify the design specification in terms of use-cases, requirements and claims. *Use-cases* are short and structured prototypical examples of the envisioned use of the system. *Requirements* describe what the machine should be able to do in order to make the use cases possible. *Claims* described the expected advantages and disadvantages of the requirement and are used to justify the requirements by describing the effect. A full description of the methodology can be found in [6].

In co-evolution with the use cases, requirements and claims, ontologies are developed. In the design phase, the ontologies are simply lists of definitions of terms occurring in the requirements, thereby making the requirement statements more precise. Later, in the implementation phase, the ontologies form the basis of knowledge representation and specify the shared vocabulary with which different system components interact.

Below, we will further describe the content-based sCE method for AmI applications. We distinguish between the generic design specification specifying the universal principles behind an (RM)^3 application, and the case specific design specification which describe specific applications.

2.1 Generic (RM)³ Design Specification

User requirements can be specified at different levels of abstraction. We have developed a generic tree of high-level user requirements, that specify the universal ideas behind (RM)³ and can be reused across various applications. The tree is presented below as a screenshot from the requirements management tool SCET [5] which we have developed to facilitate reuse and collaborative development of (RM)³ design specifications.

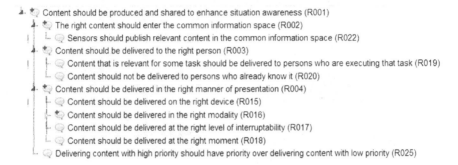

Fig. 1. Generic tree of high-level user requirements using SCET [5]

The tree of universal (RM)³ requirements is supported by a set of top-level ontologies which are specified in protégé [7].:

- *Upper content ontology* is used to describe each type of content and contains classes such as *message, sensor-reading, question*, etc.
- *Upper user ontology* is used to describe users and contains classes such as *actor* and *robot,* and contains attributes such as *hasLocation, isPerformingActivity.*
- *Upper device ontology* is used to describe devices and contains classes like *laptop, tablet, large-information-display, means-of-interaction (MOI).* It contains attributes like *supports-modality, hasInterruptiveness, isUsedBy.*

2.2 Developing Application Specific Design Specifications

The generic design specification provides a nice starting point for deriving application specific requirements and ontologies. This is usually done by specifying a use case which contains an action sequence. A snippet from an action sequence is presented below:

1. The UGV has found a victim
2. The UGV adds a Victim Form to the common information space
3. The Mission Commander who is not that busy at the moment gets the Victim Form delivered in an obtrusive manner.

Given such an action sequence, the ontologies can be extended rather straightforwardly. For example, the class *robot* is subtyped with *UGV* and *UAV.* Furthermore, requirements for delivering information to the right person can be added to the generic requirements tree. In this way, a fully specified design specification emerges which are grounded in a set of coherent ontologies. After this is finished, the implementation phase begins.

3   Agent-Based Framework

The purpose of the agent-based framework is to deliver new messages according to the (RM)[3] philosophy. A high level overview of this framework is presented in figure 2.

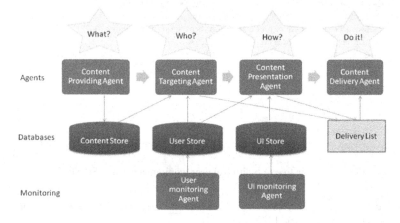

Fig. 2. Agent-based framework

The monitoring agents, presented at the bottom of Fig. 2, collect information about the user. The ontology-based databases, presented at the middle of Fig. 2, store relevant information about content, users and their user interfaces. The agents presented at the top of Fig.2 are responsible for routing information in an (RM)[3] way. We will discuss the different components in further depth below.

The *content providing agent* can be a sensor or an electronic form which gathers information from its user. The agent formats the information according to the ontology, adds metadata (e.g. source, creation time, creation GPS location) and stores it in the *content store*. The set of content providing agents determine *what* information will be shareable within the (RM)[3] system.

The *content targeting agent* decides for each piece of new information to *whom it* must be sent and with which priority. The relevance of information is determined by comparing the metadata of the content with the metadata of the user. For example, if a message is known to be created at a certain location, and one of the users is currently at that location, the agent might decide that the message should be sent to that user. Multiple targeting agents can be implemented, each specializing in specific content types. Information about the user is constantly updated by the *user monitoring agent* which stores this information in the *user store*. The monitoring agent could be very sophisticated and take various sensor data into account, such as GPS and physiological data. A more simple solution could be a monitoring agent that consults the user's agenda or simply prompts the user with a question which task it is performing.

The role of the *content presentation agent* is to decide *how* (in which form and in which modality) information must be presented to a user. This is done by taking into account the priority of the message, what the user is doing (represented in the *user store*) and what user interfaces are available to reach the user (represented in the *UI store,* such as a PDA, headphone or tactile vest). For example, if the user is doing a task which requires full visual attention, this agent might decide that the information should be presented through headphone instead of the PDA. Also, if a message is represented with low priority, the agent might decide that the information should be made available for information pull, e.g. in the user's inbox of an email-like application. The *UI Monitoring agent* keeps an overview of which user interfaces are available to the user. It monitors which means of interaction (devices and applications) are available to the user, and which content types are supported by these means of interaction.

At the end of the chain is the *content delivery agent.* It monitors the delivery list for delivery items with a specification of content, relevance and means of interaction (application + device). It will try to display the content in the delivery items, most relevant items first. The receiving application handles the piece of information and notifies the content delivery agent that the item has been received so it can update the delivery list.

The necessary information exchange between all agents in the system requires that content, applications and devices are specified in uniform ontologies. These ontologies should remain as simple as possible, yet be sufficiently expressive to provide enough information to for the agents to decide what is (RM)[3].

4 USAR Demonstrator

This section describes the design, use and first evaluation of the framework for an USAR application.

4.1 USAR Prototype

We applied our framework to an application for managing tactical information which is collected by humans and robots in an Urban Search and Rescue scenario (used in the NIFTi project [9]). The scenario consisted of a chemical freight train accident with hazardous materials and human victims. The goal of the team was to rescue the victims, build up awareness of the environment, situation and stay unharmed themselves. The participating team members were:

- A Mission Commander whose task was to get an overview of the situation and make decisions.
- An In-Field Rescuer. His role was to perform Triage, rescue victims, Support the UGV when necessary and stay unharmed.
- A UGV Operator. Controlled the UGV on macro level and if necessary on micro level. Took pictures of the environment with the Robot sensors.
- A UGV (Unmanned Ground Vehicle) The UGV is capable of performing a number of intelligent behaviors such as autonomous navigation, detecting objects (s.a. cars, victims). Its task was to go into the hot zones where a humans cannot go because of e.g. hazardous materials.

Within this scenario we used a Common Operational Picture application called TrexCOP, an application that demonstrates the $(RM)^3$ philosophy. It is able to display the information through two Means of Interactions: unobtrusive as an icon on a Map for non-important messages or obtrusive with a Pop-up Message for important messages. The information that was generated during the experiment was routed by $(RM)^3$ Agents and stored in ontology-based content stores (using OWLIM [10]).

The evaluation took place at the training center of the Fire brigade Dortmund and the participants were all firemen (professional and voluntary). During one evaluation session there were two participants, and in total there were five evaluation sessions. The firemen either fulfilled the role of in-field rescuer or UGV operator. The rescuers in the field where equipped with TrexCop on a Windows Tablet and the "desk" rescuers used TrexCop on a laptop.

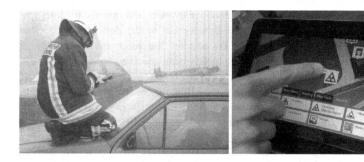

Fig. 3. Left, in-field rescuer with TrexCOP; righ, screenshot of TrexCOP

4.2   Results

In this section the results are described of the feedback from participants and the $(RM)^3$ agents log files. The Information flows through the system through various $(RM)^3$ agents. First the *Content Providing Agent* adds the data, enriched with meta data, to the system. For example, the in-field rescuer submits a victim-report with a certain location. Second the *Content Targeting Agent* calculates the priority of each piece of information for each receiver (Recv) and adds the corresponding DeliveryItem (DI) to the delivery list. This set of DeliveryItems is input for the third agent, the *Content Presentation Agents*. It decides which DeliveryItem is send to the user through which Means of Interaction (MOI). For example, a medic who is close to the victim is given the in-field-rescuer's report in an obtrusive way. See figure below for an excerpt of its log files. The fourth agent is coupled with an MOI and presents the piece of information to the user.

```
[10:43:31] Found new DI: <VictimF37, Recv: person_1_ugvop1, Prio: 0,5079, MOI: >
[10:43:31] Found new DI: <VictimF37, Recv: person_2_miscom, Prio: 0,9069, MOI: >
[10:43:31] Found new DI: <VictimF37, Recv: person_3_infieldres1, Prio: 0,5046, MOI: >
[10:43:31] Changed DI: <VictimF37, Recv: person_2_miscom, Prio: 0,9069, MOI: TrexCOP-Map-44>
[10:43:31] Created New DI: <VictimF37, Recv: person_2_miscom, Prio: 0,9069, MOI: TrexCOP-PopUp-44>
[10:43:31] Changed DI: <VictimF37, Recv: person_3_infieldres1, Prio: 0,5046, MOI: TrexCOP-Map-64>
```

Fig. 4. Screen shot of the Content Presentation Agent logs

One piece of information was handled by the Content Presentation Agent in the above log excerpt. It found three delivery items for the three actors in the scenario. For the Mission Commander the priority was high so it sent the information to the corresponding TrexCOP-Map and TrexCOP-PopUp Means of Interaction. For the In-Field Rescuer the priority was not that high so it routed it only to the TrexCOP-Map Means of Interaction. The UGV Operator was not connected to the system. So the presentation could not deliver this piece of information yet.

During the evaluation we observed the participants and they also gave feedback to the test leader. One of the results was that especially the in-field rescuer liked to add to the common picture with a team and by use of a tablet in the field. Another result was that the calculation of the priority of the information for a specific actor should be improved further. An example of this was that the in-field rescuer found it confusing that some objects that were added by himself, were not shown on his Trex-COP.

After the evaluation, the participants filled out several questionnaires. The most important feedback was that the messages that were pushed over the current screen were appreciated but were too much interfering with their current task. This issue will lead to a refinement of requirements in the next development cycle.

5   Conclusion

Optimal support of professionals in complex ambient task environments requires a system that delivers the Right Message at the Right Moment in the Right Modality (RM^3). We showed how a content and ontology-based approach can be used to cope with the inherent design complexity of such systems. We tested our approach in a real-world setting where humans, robots, collaborate together to rescue victims and exchange information using a tactical information display which is available on mobile tablet displays and fixed laptops. The first results are promising, and we intend to apply the approach to more complex situations (i.e. more actors, devices, types of information) in the future.

References

[1] van Diggelen, J., van Drimmelen, K., Heuvelink, A., Kerbusch, P.J., Neerincx, M.A., van Trijp, S., van der Vecht, B.: Mutual empowerment in mobile soldier support. Journal of Battlefield Technology 15(1), 11 (2012)

[2] Chung, L., do Prado Leite, J.C.S.: On non-functional requirements in software engineering. In: Borgida, A.T., Chaudhri, V.K., Giorgini, P., Yu, E.S. (eds.) Conceptual Modeling: Foundations and Applications. LNCS, vol. 5600, pp. 363–379. Springer, Heidelberg (2009)

[3] Luyten, K., Van den Bergh, J., et al.: Designing distributed user interfaces for ambient intelligent environments using models and simulations. Computers & Graphics 30(5), 702–713 (2006)

[4] Gross, T.: Towards a new human-centred computing methodology for cooperative ambient intelligence. Journal of Ambient Intelligence and Humanized Computing 1(1), 31–42 (2010)

[5] situated Cognitive Engineering Tool, http://tno.scetool.nl/

[6] Neerincx, M.A., Lindenberg, J.: Situated cognitive engineering for complex task environments. In: Schraagen, J.M.C., Militello, L., Ormerod, T., Lipshitz, R. (eds.) Naturalistic Decision Making & Macrocognition, vol. 390, pp. 373–390. Ashgate Publishing Limited, Aldershot (2008)

[7] Protégé ontology editor, http://protege.stanford.edu/

[8] Abowd, G.D., Dey, A.K.: Towards a better understanding of context and context-awareness. In: Gellersen, H.-W. (ed.) HUC 1999. LNCS, vol. 1707, pp. 304–307. Springer, Heidelberg (1999)

[9] Kruijff, G., Colas, F., Svoboda, T., van Diggelen, J., Balmer, P., Pirri, F., Worst, R.: Designing Intelligent Robots for Human-Robot Teaming in Urban Search & Rescue. In: Proceedings of the AAAI 2012 Spring Symposium on Designing Intelligent Robots (2012)

[10] Kiryakov, A., Ognyanov, D., Manov, D.: OWLIM – A Pragmatic Semantic Repository for OWL. In: Dean, M., Guo, Y., Jun, W., Kaschek, R., Krishnaswamy, S., Pan, Z., Sheng, Q.Z. (eds.) WISE 2005 Workshops. LNCS, vol. 3807, pp. 182–192. Springer, Heidelberg (2005)

Tracking People and Equipment Simulation inside Healthcare Units

Cátia Salgado, Luciana Cardoso, Pedro Gonçalves,
António Abelha, and José Machado*

Universidade do Minho, CCTC, Braga, Portugal
{a56840,a55524}@uminho.pt,
{pgoncalves,abelha,jmac}@di.uminho.pt

Abstract. Simulating the trajectory of a patient, health professional or medical equipment can have diverse advantages in a healthcare environment. Many hospitals choose and to rely on RFID tracking systems to avoid the theft or loss of equipment, reduce the time spent looking for equipment, finding missing patients or staff, and issuing warnings about personnel access to unauthorized areas. The ability to successfully simulate the trajectory of an entity is very important to replicate what happens in RFID embedded systems. Testing and optimizing in a simulated environment, which replicates actual conditions, prevent accidents that may occur in a real environment. Trajectory prediction is a software approach which provides, in real time, the set of sensors that can be deactivated to reduce power consumption and thereby increase the system's lifetime. Hence, the system proposed here aims to integrate the aforementioned strategies - simulation and prediction. It constitutes an intelligent tracking simulation system able to simulate and predict an entity's trajectory in an area fitted with RFID sensors. The system uses a Data Mining algorithm, designated SK-Means, to discover object movement patterns through historical trajectory data.

Keywords: RFID object tracking, Trajectory prediction, Simulation, Healthcare, Data Mining, SK-Means.

1 Introduction

The Radio-Frequency Identification (RFID) is a tecnology that allows the development of object tracking systems, in order to guarantee an improvement on safety, healthcare quality and a reduction of costs. Today, most of the tracking systems implemented in hospitals are based in this techology [1,2,3,4].

It is proved that these systems provide a reduction on the time spent by nurses and staff localizing medical equipment, hence improving productivity. On the other hand, there is a significance reduction on costs associated to equipments theft, loss and misplacing, leading to greater utilization. Adopting an RFID

---

* Corresponding author.

A. van Berlo et al. (Eds.): *Ambient Intelligence – Software & Applications*, AISC 219, pp. 9–16.
DOI: 10.1007/978-3-319-00566-9_2   © Springer International Publishing Switzerland 2013

based system to prevent these incidents becomes a potential solution for those hospitals that spend millions on stolen or lost equipments every year [5,6,1].

Recently, the topic of saving energy in sensor networks has attracted much attention. Being able to correctly predict an entity trajectory is one approach to be considered when dealing with this matter. In fact, an accurate trajectory prediction in sensor networks can benefit energy saving as it is possible to find out which sensor nodes will not detect the object being tracked, thus reducing its utilization or inactivating it. Most people make trajectories of a repetitive nature, making it possible to predict the next route of a person based on his previous history. The topic of trajectory prevision can be treated in a simulating context, and a person's trajectory can be easily obtained based on his personal route prediction [7,8,9,10,11,12,13,14].

Create a simulation environment it is essential to test and evaluate the system. With this simulator, we can correct existing errors, test changes in the tracking system without costs and prevent accidents that occur in test environment. This system can help the study of the system behavior when subjected to specific challenges.

One main purpose of this study is to evaluate the influence of the RFID sensors on predicting the trajectory of the object. Does it make any difference to place sensors all along the path, or to just keep some at specific points? Secondly, what about its range? Is it worth the investment on a better sensor, with a higher range of detection? What are the critical points on this floor that demand RFID sensors? All these questions can be addressed in a simulation context, and the hospital shall benefit from a study of these aspects in the proposed simulator. In this paper some examples will be presented to illustrate how the hospital can use the simulator to solve this issues.

2   Data Mining

A clustering Data Mining algorithm, designated SK-Means, was implemented to calculate those points where the object is most probably going to pass based on the history of trajectories. The chosen algorithm is well-known for its simplicity and is frequently used for discovering movement patterns [10,15]. SK-Means can only be used if the number of clusters is previously known [10,15,16]. Thus, a dataset containing all the movement sizes is maintained in order to give an approximate measure of the number of clusters needed. Since ones wants a prediction path similar to the real path this number is an average of the registered movement sizes. Also, the travelled path is stored in a structured dataset to be used in subsequent iterations. Before starting the movement, the simulator analyses the dataset, containing all previous locations, and extracts the information it needs to accomplish the purpose of calculating an estimated path. With a large number of previous point locations and a predefined number of clusters, the algorithm is capable of determining a set of coordinates, enclosing the entire area, which represents the initial predicted path.

3   Implementation

The proposed system is based on Java programming language, which provides robustness and scalability through a set of APIs such as Swing and Weka, both essential for the system's design and Data Mining respectively.

In order to maintain a full record, it was used a MySql database, which guarantees security, reliability, trust and an easy access to the stored data.

The Pentaho Report Designer is an open source application that provides a visual design environment to create reports and an easy integration of external data from databases such as MySql. It was used to create dynamic statistics of the results obtained in different scenarios.

The points travelled by the object are stored in a structured dataset at the end of the movement, along with the size of that trajectory. As mentioned before, to proceed with the SK-Means cluster analyzes, the number of clusters will be calculated in advance. Since we want a prediction path similar to the real path we consider this number to be an average of the registered movement sizes, that is the reason why it is maintained a record of these values. With a large number of previous point locations and a predefined number of clusters, the algorithm is capable of determining a set of coordinates enclosing the entire area that represent the initial predicted path.

The developed system in Java, which is essentially the simulator module, relies on the interaction between the intelligent agent responsible for the learning process and the one responsible to simulate a trajectory. Gathering both information, it is possible to calculate the prediction results at the end of the movement. These results are stored in a MySql database to posterior analyzes and reporting in Weka.

4   Trajectory Simulation

Beginning with the primary work developed by Gonçalves et al.[17], several changes were applied to improve the performance of the algorithm and to adapt it to the dynamic of the chosen path. Where before we had a simple environment, with no obstacles, representing a total of 150 positions, now we have an hospital floor filled with doors, rooms, corridors and elevators, with 14506 possible positions that the object must travel in order to simulate a more accurate algorithm. New features regarding the movement on corridors were added to the trajectory concept. The Graphic User Interface is represented in Figure 1.

Figure 2 shows the simulation environment with three robots, distinguished by their color and identification, moving around on the hospital floor. The simulator allow for starting each object movement whenever the user decides to, and to program them to move at distinct velocities.

4.1   Corridors

In order to submit the object to a certain direction, each corridor is associated with a number (-1, 0 or 1) that represents the preferable direction the object will

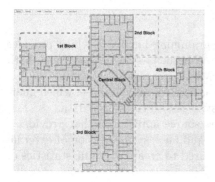

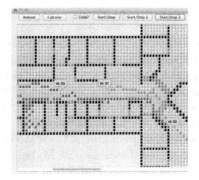

Fig. 1. Actual graphic user interface

Fig. 2. Simulating the movement of three objects simultaneously

follow in xx (xdir) and yy (ydir). So, for a given position within the corridor, xdir and ydir will be calculated under the influence of these two parameters. For example, if the values assumed for corridor x are (0,1), then xdir component will have much less effect on the movement (xdir will be much inferior) than the ydir component. In other words, the object will move preferentially on yy direction, and less on xx. The problem concerning the object being "stuck" around some place is eliminated, as it keeps on moving through the given direction. Another advantage related to this approach is that it is quite simple and intuitive to, at any time, change the course of the study and define new paths, by simply alter the values of those two parameters, and the initial and final points of the movement.

4.2   Rooms

Every time the object reaches a room, its next position is determined randomly. Random points are generated in order to create an unpredictable movement along the selected area, a method achieved by using the concept of gravitational forces. Given a random point, the algorithm will give a direction based on the gravitational force calculated between the current point location and the random one, which can be attractive or repulsive. This means that when inside a room, from the moment it enters to the moment it leaves, the object will have a complete random movement. In fact, considering the process of tracking assets inside a hospital, it is not relevant to predict the path inside the room, it is only necessary to know that the object has enter it.

5   Results

When the robot object reaches the end of the board, the results treatment module takes place. Based on the estimated and real trajectory it is possible to evaluate the model precision, and is present to the observer an intuitive perception of the similarity between both paths.

The number of correct and incorrect predictions are calculated iterating both real and estimated paths together, having both agents working cooperatively. The precision is given by:

$$\frac{correctly\ predicted}{correctly\ predicted + wrongly\ predicted} 100 \qquad (1)$$

All these values are stored in a dataset to allow a subsequent analysis, namely to calculate the precision average for each case, and detect and eliminate movement logs that do not fit the pattern.

Four different scenarios were studied in order to evaluate the simulation dynamics and the prediction results under different circumstances. The first study case (Figure 3) has to do with an area with no sensors available, except those on elevators. The predicted path is distributed all along the floor's 5 blocks of circulation at any time, as there are no sensors limiting the possible object location. In the second scenario (Figure 4), sensors were placed at each room's door covering the entire width of entry or exit, thus preventing the object passes without being captured. The estimated path is restricted to a certain area, accordingly to the last object position captured by sensors. So, if a sensor responsible to detect room entrances, detects that the object has entered the room, the estimated path in that moment will be the predicted points calculated by data mining enclosed on that room. Another way of reducing the object's possible location is to place sensors where the four blocks communicate with the central block (Figure 5). In this case, no predicted path will be shown for those blocks where the object never enters. Similarly, a fourth scenario (Figure 6) regards the case in which the sensors architecture replicates the real architecture deployed in the Hospital floor.

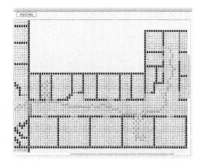
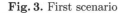

Fig. 3. First scenario

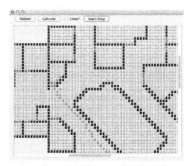

Fig. 4. Second scenario

Table 1 resumes the average precision obtained for each scenario. The prevision started with four trajectory records and proceed until the object conclude three hundred movements. Different historical trajectories were maintained for each scenario. Figure 7 illustrates the precision evolution as a function of the

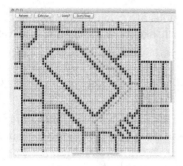
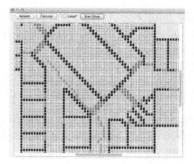

Fig. 5. Third scenario   **Fig. 6.** Fourth scenario

number of records in one of the aforementioned scenarios. This graphic shows the precision when the object describes an usual and an unusual trajectory (red and black lines, respectively). It also demonstrates that the algorithm converges quickly.

Table 1. Precision results

Scenario	1st Elevators	2nd Doors	3rd Corridors	4th Hospital
Number of sensors	4	124	10	10
Average Precision	56.4%	60.8%	60.1%	62.5%

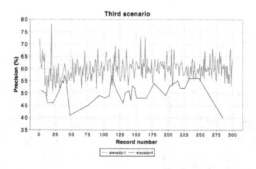

Fig. 7. Precision evolution as a function of the number of records - scenario 3

When sensors are placed at each door, in a total of 124 sensors, the algorithm is capable to predict the travelled path with a precision/accuracy of 61% (average). If the sensors are placed on corridors, instead of rooms, only 10 sensors are needed, and the value obtained in this case is similar - 60%. The best precision average recorded, although, concerns the scenario that represents the Hospital RFID sensors distribution. Here, the prevision model was able to estimate positions with a 62% precision. This value decreases to 56% when no sensors at all are available (limiting the possible object location). An higher accuracy is obtained when intelligent agents, that have adequate knowledge about one another, are used to retrieve predicted locations based on reality.

6   Conclusions and Future Work

The simulator is now capable of predicting behaviours in an ambient more realistic than before [17]. The object trajectory has been improved to meet the requirements of a healthcare environment. It can move through different areas, visit rooms and follow predefined directions chosen by the programmer. It is possible to define any desirable path within the hospital's 6th floor and initiate the prediction system based on that behavior. The prediction system works for any initial and final points the user may want to apply, by simply changing the directions associated with corridors. Also, it is now possible to simultaneously visualize the predicted and simulated path in real time.

The trajectory algorithm has a new feature that, when selected, prevents the object from going more than once through the same point which resolves the problem related with statistical computation.

One of the major achieved improvements was the insertion of multiple entities in the simulation environment at the same time. The proposed simulator is capable of showing many moving objects simultaneously, which constitutes the fulfilment of a hospital requirement. Indeed, it is anticipated that this system will be incorporated as a decision-making support tool in hospitals in the future.

Introducing the walls, rooms, corridors and doors coordinates into the simulator was one of the most time consuming parts of this work. In order to streamline the simulator we propose as future work an application focused on an automatic reading of images (regarding areas the user may intend to simulate) and a subsequent automatic input of these entities coordinates into Java language.

The predicted path should be calculated in real time, accordingly to the object trajectory and the positions captured by sensors. Indeed, it would be more efficient and logic to add points to the predicted path as the object passes through sensors, instead of eliminating those that do not belong.

As future work, we advised the hospital to adopt a strategy of energy saving based on the prevision model. We anticipate that it is possible adapt the system easily, in order to retrieve the sensors that can be unactivated and the period that should be asleep until they are needed again.

Acknowledgements. This work is financed with the support of the Portuguese Foundation for Science and Technology (FCT), with the grant SFRH/BD/70549/2010 and within project PEst-OE/EEI/UI0752/2011.

References

1. Qu, X., Simpson, L.T., Stanfield, P.: A model for quantifying the value of rfid-enabled equipment tracking in hospitals. Advanced Engineering Informatics 25(1), 23–31 (2011)

2. Wyld, D.C.: Preventing the worst case scenario: An analysis of rfid technology and infant protection in hospitals. The Internet Journal of Healthcare Administration 10(1) (2010)
3. Oztekin, A., Pajouh, F.M., Delen, D., Swim, L.K.: An rfid network design methodology for asset tracking in healthcare. Decision Support Systems 49(1), 100–109 (2010)
4. Laskowski, M., Demianyk, B., Naigeboren, G., Podaima, B., Friesen, M., McLeod, R.: Rfid modeling in healthcare. Sustainable Radio Frequency Identification Solutions, 217–240 (2010)
5. Macalanda, E.C.: Radio frequency identification (rfid) for naval medical treatment facilities (mtf). Master's thesis, NAVAL POSTGRADUATE SCHOOL (September 2006)
6. Fuhrer, P., Guinard, D.: Building a smart hospital using rfid technologies
7. Lin, K.W., Hsieh, M.H., Tseng, V.S.: A novel prediction-based strategy for object tracking in sensor networks by mining seamless temporal movement patterns. Expert Systems with Applications 37(4), 2799–2807 (2010)
8. Tseng, V.S., Lu, E.H.C.: Energy-efficient real-time object tracking in multi-level sensor networks by mining and predicting movement patterns. The Journal of Systems and Software 82(4), 697–706 (2009)
9. Hsu, J.M., Chen, C.C., Li, C.C.: Poot: An efficient object tracking strategy based on short-term optimistic predictions for face-structured sensor networks. Computers and Mathematics with Applications 63(2), 391–406 (2012)
10. Chen, L., Lv, M., Chen, G.: A system for destination and future route prediction based on trajectory mining. Pervasive and Mobile Computing 6(6), 657–676 (2010)
11. Tseng, V.S., Lin, K.W.: Energy efficient strategies for object tracking in sensor networks: A data mining approach. The Journal of Systems and Software 80(10), 1678–1698 (2007)
12. Anagnostopoulos, T., Anagnostopoulos, C.B., Hadjiefthymiades, S., Kalousis, A., Kyriakakos, M.: Path prediction through data mining. International Conference on Pervasive Services, 128–135 (2007)
13. Chen, L., Lv, M., Ye, Q., Chen, G., Woodward, J.: A personal route prediction system based on trajectory data mining. Information Sciences 181(7), 1264–1284 (2011)
14. Colak, I., Sagiroglu, S., Yesilbudak, M.: Data mining and wind power prediction: A literature review. Renewable Energy 46, 241–247 (2012)
15. Lam, Y.K., Tsang, P.W.M.: exploratory k-means: A new simple and efficient algorithm for gene clustering. Applied Soft Computing 12(3), 1149–1157 (2012)
16. Mehta, S., Shete, D., Lingayat, N., Chouhan, V.: K-means algorithm for the detection and delineation of qrs-complexes in electrocardiogram 31, 48–54 (2010)
17. Gonçalves, P., Alves, L., Sá, T., Quintas, C., Miranda, M., Abelha, A., Machado, J.: Object trajectory simulation - an evolutionary approach. In: Novais, P., Machado, J., Rodrigues, C., Abelha, A. (eds.) Modelling and Simulation, EUROSIS (2011)

Requirements Systematization through Pattern Application in Ubiquitous Systems

Tomás Ruiz-López, Manuel Noguera,
María José Rodríguez Fórtiz, and José Luis Garrido

Software Engineering Department, University of Granada
Periodista Daniel Saucedo Aranda s/n, 18.014 Granada, Spain
{tomruiz,mnoguera,mjfortiz,jgarrido}@ugr.es

Abstract. Application of patterns to address Non-Functional Requirements (NFRs) is a field that has not been widely explored, and that is
still uncovered in the scope of Ubiquitous Computing. The unique features of this paradigm, such as context-awareness or technological unobtrusiveness, present a challenge to appropriately treat the specific NFRs
related to this field. In this paper, recurring situations in ubiquitous systems have been identified and captured as patterns, which can be used
to satisfy NFRs in different domains.

Keywords: requirements engineering, patterns, ubiquitous computing,
ambient intelligence.

1 Introduction

Ubiquitous Computing features are tightly related to the satisfaction of certain
Non-Functional Requirements (NFRs) [2]. For instance, this paradigm claims
that technology should remain invisible, providing *unobtrusive* access to the
user, or that systems should be able to *adapt* themselves based on contextual
changes.

Moreover, there are recurring situations that happen in most ubiquitous systems where there is a direct impact to the satisfaction of NFRs. That is the case,
for example, of providing input/output mechanisms that are aware of contextual
circumstances. *Usability* of the system is partially determined by the assessment
of the I/O mechanisms.

Failure to systematically treat NFRs may result on a loss of the overall quality
of the system. To overcome this problem, we proposed a method for Requirements Engineering (REUBI) [12] that aims to systematically deal with NFRs.
However, the provided support is broad and it does not capture recurring situations and their treatment to maintain the system quality.

In this regard, a set of requirements patterns [13] have been identified. This
allows to provide reusable, well-known solutions to common problems, enhancing
the satisfaction of the NFRs that are related to the patterns.

A. van Berlo et al. (Eds.): *Ambient Intelligence – Software & Applications*, AISC 219, pp. 17–24.
DOI: 10.1007/978-3-319-00566-9_3   © Springer International Publishing Switzerland 2013

The rest of this paper is structured as follows. Section 2 presents a set of requirements patterns for ubiquitous systems, organized in five categories: decomposition, resolution, operationalization, selection and ethics. The impact of the application on the NFRs is studied in Section 3. Finally, the main conclusions and future work are outlined in Section 4.

2   Requirements Engineering Patterns for Ubiquitous Systems

Requirements analysts have to face similar situations during the requirements stage across different projects for Ubiquitous Environments, no matter what their scope is. Being able to reuse previous knowledge about other projects may be helpful in order to provide a more systematic treatment of the requirements, and consequently, reducing the time taken to perform this activity.

Patterns are good means to achieve this goal. They capture recurring situations in a well documented form, so that they can be applied regardless the intent of the final system and communicated between different analysts. In this paper is presented a set of patterns for ubiquitous systems that have been identified from the analysis of existing systems, the authors' experience in the development of these systems, and the adaptation of existing design patterns to the requirements engineering stage [4]. They have been organized into five categories: decomposition, resolution, operationalization, selection and ethics. This organization responds to the application of these patterns in the corresponding stages of REUBI. To the best of our knowledge, no previous work has been carried out in the Ubiquitous Computing field regarding NFRs. Table 1 summarizes the intent of each pattern. They are described in the following sections.

2.1   Decomposition Patterns

This set of patterns focus on the decomposition of top level *goals* (i.e., objectives that need to be addressed) into more concrete, finer-grained ones which can be *operationalized* (i.e., looking for architectural/design decisions to contribute to the satisfaction of the goals). Existing approaches suggest to apply *type* or *topic* based decompositions to divide a goal. Also, decomposition catalogs can be employed to reuse recurrent decompositions.

The patterns presented in this section focus on characteristics that are essential in ubiquitous systems. In particular, the *Ubiquity definition* pattern proposes the incorporation of an objective, titled *Ubiquity*. In order to be able to define what ubiquity means in the context of the system under analysis, this objective should be further decomposed into more concrete sub-objectives; for instance, taking into account the main features of the paradigm, it could be decomposed into *Enhance invisibility of technology*, *Create smart spaces*, or *Mask uneven conditioning* [14]. Precisely defining the concept of ubiquity in the scope of the project helps to make decisions in order to facilitate the design of the required system.

Table 1. Summary of requirements patterns grouped into different categories

Category	Pattern	Description
Decomposition	*Ubiquity definition*	To be able to clearly specify the notion of ubiquity in the system under analysis, so that operationalizations can be found to satisfy this requirement.
	Context-narrowing decomposition	To be able to break down objectives into finer-grained ones taking into account the structure of the context situations that they have to be satisfied under.
Resolution	*Obstacle-objective chain*	To establish an obstacle resolution strategy taking into account the likelihood of occurrence of the obstacles.
	Severity-driven resolution	To be able to modify the obstacle resolution strategy if the severity of an obstacle changes under certain contextual circumstances.
Operationalization	*Context-sensitive I/O*	To determine the input/output methods that should be used taking into account context, in order to maintain the satisfaction of other Non-Functional Requirements.
	Unobtrusive identification	To be able to perform security-sensitive actions in a secure way without interfering the usual execution of those actions or bothering the user.
	Redundant operationalization	To avoid a single point of failure that can cause the failure to fulfill a Non-Functional Requirement.
Selection	*Variable priority*	To specify the prioritization changes in an objective depending on changes in context attributes.
Ethics	*Pseudonymity*	To set up the amount of user's personal data that is exchanged depending on the trustworthiness of the context situation.
	Human factor	To be able to analyze which parts of the system are performed in a better way by a human agent instead of pointlessly introducing technology to perform them.

As mentioned above, objectives need to be decomposed. The *Context-narrowing* pattern proposes to break down objectives taking into account the structure of the context. The realization of a goal may differ depending on contextual variables; therefore, splitting it into different sub-goals which apply under certain context situations helps to address their realization in a more suitable way [9].

2.2   Resolution Patterns

Effective *obstacle* (undesirable situations that can hinder the satisfaction of an objective) overcoming has two facets: prevention and resolution. This set of patterns aims to provide strategies to be followed to identify and solve problems caused by the appearance of inconvenient situations that may hinder system goals.

Likelihood of obstacle occurrence is an important factor in order to determine the protocol to follow, both to prevent and correct possible misbehaviors. The *Obstacle-objective chain* pattern proposes to sort the obstacles according to their likelihood, and find objectives to mitigate them accordingly. Probabilities can be difficult to estimate; however, a rough estimation based on the frequency of occurrence of inconvenient situations may be sufficient [1].

Another important factor is the severity of the obstacle, if it happens. Moreover, severity may change under certain contextual circumstances. The *Severity-driven resolution* pattern extends the previous one taking into account the severity of the obstacle to define alternative resolution protocols when the severity changes depending on context. It is important to note that likelihood of occurrence may also change upon contextual constraints, since severe obstacles have lower probability to occur.

2.3   Operationalization Patterns

Searching the solution space to find the most suitable technology or method in Ubiquitous Computing is a hard task given the heterogeneity of available solutions and the different quality properties that they present. Moreover, there is *no silver bullet*; none of the existing solutions can give the best results in every situation. Thus, adaptation can be a good choice to overcome this problem.

That is the case of the *Context-sensitive I/O* pattern. It states that different input/output mechanisms need to be applied under different contextual circumstances in order to avoid disruptions in the user's activities [4]. The interaction mechanisms should make use of natural interfaces in order to be unobtrusive and easy to use. Therefore, the system needs to adapt to provide the most suitable interaction mechanism to be applied depending on the context situation.

Similarly, there are some activities that deal with security-sensitive data from the users. Software systems need to authenticate users in order to give access to this information of operations. The *Unobtrusive identification* pattern proposes to apply different mechanisms in order to authenticate users in an unobtrusive way [6]. To do that, actions need to be sorted according to their security level. Then, for each of them, analysts should find security mechanisms that are least invasive for the users, but at the same time, guarantee secured access to resources.

In other occasions, a usual problem that may happen is the failure of one of the operationalizations that are responsible for the satisfaction of an objective. This is the case of, for instance, sensors or actuators. There are many possible scenarios where their functioning is not as expected, e.g., due to a power outage or a hardware failure. The *Redundant operationalization* pattern proposes the introduction of a combination of operationalizations [12]. Redundancy is a well-known technique in distributed systems in order to increase the robustness of the system in case of failure of one of its parts. It leads to avoid a single point of failure. Grouping and combining different alternatives to satisfy a goal also triggers the appearance of new properties that the operationalizations do not have standalone, both positive and negative. Also, the interactions among operationalizations should be studied in order to determine if they can be applied simultaneously.

2.4   Selection Patterns

Selection among different alternatives in the REUBI method is done following a quality-driven approach; i.e., the satisfaction of NFRs usually determine the election or not of a certain operationalization. When a tradeoff happens, i.e. a conflicting situation between two NFRs, priority of the NFRs is used to break the tie [10].

However, since priority may vary under different situations, this entails that different selections can be done. The *Variable priority* pattern aims to capture this situation. For each priority level that can be assigned to a requirement given a certain context, the analyst needs to find possible operationalizations that address this change. Note that different operationalizations can be possible for different prioritization levels (critical, important or normal), but there can be cases with the same operationalizations.

2.5   Ethics Patterns

Ubiquitous Computing is user-centric in nature. This means that the whole Requirements Engineering process has to be focused on fulfilling the user's needs, applying those techniques that are more suitable to unveil what the requirements for the system are. Moreover, it also entails that those NFRs which are closely related to the user, such as privacy and safety, have to be carefully treated.

Thus, to enhance the privacy of the user, the *Pseudonymity* pattern proposes the creation of a hierarchical structure with different amounts of information. The hierarchy contains all user's private data [5]. The top node of the hierarchy provides all the user's information, whereas the bottom node provides no information about him/her. Between both extremes, different amounts of information can be shared. When user's information is requested, a node of this hierarchy is selected taking into account the trustworthiness of the situation. Thus, the user decides to share certain pieces of data depending on context variables, expressing this decision as configurable user preferences that are checked when information needs to be exchanged.

Finally, current developments mainly focus on the technological aspects. Few or no works study the possibility of not introducing technology to perform certain tasks, and rather, leave a human agent perform it as usual. User's well-being can be achieved with the introduction of assistive technologies, but also by leaving them to do things on their own.

The *Human factor* pattern proposes a simple, yet important idea: the incorporation of an extra "operationalization", the *human action*. This operationalization has to be related to those goals that can be performed by a person, as well as to the softgoals, in order to study the benefits of the realization of the objectives by the technology versus the realization by a human.

There can be both positive and negative contributions to the satisfaction of the NFRs, but the evaluation procedure described by the REUBI method should be able to tell which options are ultimately the best ones to satisfy them. This evaluation procedure enables to obtain a set of operationalizations that has a better fulfillment of the requirements. Moreover, since objective prioritization may change, a hybrid approach could be the most suitable option: depending on the context, some activities would be performed by the user, and some others by the software.

3   NFRs Impacted by the Application of Patterns

The main criteria followed to the discovery of the proposed patterns is the improvement of the satisfaction of the NFRs. In fact, their application enhances the satisfaction of some of them, besides providing a more systematic way to perform requirements engineering.

Table 2. NFRs affected by the application of the proposed Requirements Engineering patterns

Non-Funtional Requirement	Ubiquity definition	Context-narrowing decomp.	Obstacle-objective chain	Severity-driven resolution	Context-sensitive I/O	Unobtrusive identification	Redundant operationalization	Variable priority	Pseudonymity	Human factor
Invisibility	✓		✓							
Unobtrusiveness	✓		✓		✓	✓		✓		
User acceptance	✓		✓	✓	✓		✓			✓
Usability	✓	✓			✓					✓
Security	✓		✓		✓				✓	
Accuracy	✓				✓	✓		✓		
Ease of development		✓			✓					
Ease of maintenance		✓			✓					
Development time and cost	✓	✓	✓	✓	✓	✓	✓	✓		
Performance		✓				✓				
Robustness			✓	✓		✓				
Privacy						✓			✓	✓
Safety			✓			✓				
Adaptability					✓	✓	✓	✓	✓	
User's autonomy									✓	✓
Interface compatibility					✓	✓	✓		✓	
Reliability			✓	✓			✓			
Recovery			✓	✓						

Table 2 summarizes the interactions that arise between NFRs and each of the proposed patterns. These interactions can be both positive and negative, depending on the actual project under analysis. Consider, for instance, *Development time and cost*. The application of some of these patterns may involve the need to develop extra components, as well as adaptation mechanisms; e.g. to determine which input/output method should be used in a certain context situation, as proposed in the *Context-sensitive I/O*. These effects can be mitigated by the application of additional development techniques, like the reuse of components or services in a Service-Oriented Architecture (SOA) [3] or the application of Model-driven Development (MDD) techniques [11].

Some other requirements are easily improved. That is the case of those NFRs related to the development process. The application of patterns constitute a source of reusable assets that can be applied across different projects with the introduction of some small modifications to better fit the project needs. The adaptability of the software can be also improved, since adaptations are the main focus of some of the patterns.

Last, but not least, the satisfaction of some other NFRs can be enhanced, but it depends on the final instantiation of the patterns by the analysts. Special emphasis should be made in those requirements related to the user experience. Unobtrusiveness, usability, accuracy or security, among others, are key requirements that can lead to the acceptance of the system if they are treated in an appropriate way.

4   Conclusions and Future Work

The patterns presented in this paper stem from the analysis of ubiquitous systems and are the core of a more extensive set of patterns that are being identified. Many examples of their application can be found on the bibliography, where they have proven to be useful. Nevertheless, it is still necessary to perform an exhaustive evaluation of the suitability and usefulness of these patterns in the requirements engineering phase.

Patterns have several advantages: they speed up the development, increase reusability of software assets, provide general solutions that can be applied in several fields, and provide a common terminology to refer to recurring problems, improving communication among analysts and designers. However, they should be carefully applied, since a misuse of pattern application may lead to an unnecessary increase of software complexity [7].

Requirements patterns may be useful both to satisfy certain NFRs, specially some that need to be carefully addressed in ubiquitous systems, and to derive a design that actually meets the requirements. This paper has briefly presented an approach to proceed to design realization after requirements patterns have been applied, but a thorough study needs to be performed in this regard to finally obtain a high quality design.

In the future, we are planning on studying the implications of these patterns in the software design phase. The aim is to take into account the patterns in a model-driven engineering approach [11] [8] to semi-automatically derive design and implementation models for ubiquitous systems.

Acknowledgements. This research work is funded by the Spanish Ministry of Economy and Competitiveness (project reference number TIN2012-38600), by the Innovation Office from the Andalusian Government through project TIN-6600, CEI BioTIC Granada under project 20F2/36, and The Spanish Ministry of Education, Culture and Sports through the FPU Scholarship.

References

1. Cheng, B., Sawyer, P., Bencomo, N., Whittle, J.: A goal-based modeling approach to develop requirements of an adaptive system with environmental uncertainty. In: Schürr, A., Selic, B. (eds.) MODELS 2009. LNCS, vol. 5795, pp. 468–483. Springer, Heidelberg (2009)
2. Chung, L., Nixon, B., Yu, E., Mylopoulos, J.: Non-Functional Requirements in Software Engineering. Springer (2000)
3. Erl, T.: SOA: Principles of Service Design. Prentice Hall (2008)
4. Landay, J.A., Borriello, G.: Design patterns for Ubiquitous Computing. IEEE Computer 36(8), 93–95 (2003)
5. Langheinrich, M.: Privacy by design - principles of privacy-aware ubiquitous systems. In: Abowd, G.D., Brumitt, B., Shafer, S. (eds.) UbiComp 2001. LNCS, vol. 2201, pp. 273–291. Springer, Heidelberg (2001)
6. Lenzini, G.: Design of architectures for proximity-aware services: Experiments in context-based authentication with subjective logic. ENTCS, vol. 236, pp. 47–64 (2009)
7. McConnell, S.: Code Complete: A Practical Handbook of Software Construction. Microsoft press (2009)
8. Object Management Group: Model Driven Architecture (2003), http://www.omg.org/mda/
9. Rossi, G., Gordillo, S., Lyardet, G.: Design patterns for context-aware adaptation. In: The 2005 Symposium on Applications and the Internet Workshops, pp. 170–173 (2005)
10. Ruiz-López, T., Rodríguez-Domínguez, C., Noguera, M., Garrido, J.L.: Towards a reusable design of a positioning system for AAL environments. In: Chessa, S., Knauth, S. (eds.) EvAAL 2011. CCIS, vol. 309, pp. 65–79. Springer, Heidelberg (2012)
11. Ruiz-López, T., Rodríguez-Domínguez, C., Noguera, M., Rodríguez, M.J.: A Model-Driven Approach to Requirements Engineering in Ubiquitous Systems. In: Ambient Intelligence-Software and Applications, pp. 85–92 (2012)
12. Ruiz-López, T., Noguera, M., Rodríguez, M.J., Garrido, J.L., Chung, L.: REUBI: A requirements engineering method for ubiquitous systems. Science of Computer Programming (2012) (in press)
13. Supakkul, S., Hill, T., Oladimeji, E.A., Chung, L.: Capturing, Organizing, and Reusing Knowledge of NFRs: An NFR Pattern Approach. In: Second International Workshop on Managing Requirements Knowledge (MARK), pp. 75–84 (2009)
14. Weiser, M.: The computer of the 21st century. Scientific American 265(3), 94–104 (1991)

Indoor Tracking Persons Using Bluetooth: A Real Experiment with Different Fingerprinting-Based Algorithms

María Rodríguez-Damián, Xosé Antón Vila Sobrino, and Leandro Rodríguez-Liñares

Dpto. Informática, Universidade de Vigo, Lagoas-Marcosende 36310, Spain
{mrdamian,anton,leandro}@uvigo.es

Abstract. In outdoor localization, global positioning systems (GPS) has been widely used. Indoor applications require a precise estimation that GPS can not achieve. Several technologies have been tried out as WI-FI, RFID, Bluetooth, Zigbee and others. This paper describes an experiment conducted in a medium-sized room in which six zones have been identified and two Bluetooth transmitters were installed. The aim is to enable continuous monitoring of areas where a person moves. For this purpose, we have used the technique of RSSI fingerprinting and tested three different algorithms. The best results were obtained with an algorithm based on SVM, which yielded success rates of 88.54%. Based on this algorithm, we intend to develop a cheap and easily configurable indoor localization system.

1 Introduction

Global Positioning System (GPS) is a well known outdoor localization and navigation system; however this technology does not work in indoor environments [1]. There have been numerous attempts at indoor localization tracking [2] but none of them have been widely adopted, primarily due to the cost of deploying and maintaining building-wide location technology [3]. Real-world location-aware applications are continuously expanding: location of products stored in a warehouse, people location in buildings, way-finding systems, etc. Different technologies have been tested in numerous systems, such as infra-red beacons in the Active Badge Localization System [4], ultrasound time of arrival in Cricket [5] or Radio Frequency Identification in LANDMARC [1].

At this moment, most of the researchers are currently focussed on two wireless technologies: Wi-fi and Bluetooth. Wi-fi has high power demands, is more difficult to set-up and maintain and is also less common in mobile phones. Instead, Bluetooth was designed for low power devices and the vast majority of the mobile phones supports it. For these reasons our research is based on Bluetooth technology [3, 6].

When using Bluetooth, RSS is a widely employed parameter, either with trilateration or with fingerprinting. Trilateration is a trigonometric approach for tracking mobile objects considering the concept of triangles. It requires a theoretical propagation model that relates RSS values with distances. Fingerprinting is the most accurate and popular method and is based on matching some characteristic of the signal that is location dependent [7, 8].

A. van Berlo et al. (Eds.): *Ambient Intelligence – Software & Applications*, AISC 219, pp. 25–32.
DOI: 10.1007/978-3-319-00566-9_4   © Springer International Publishing Switzerland 2013

We intend to design and implement a system for locating and tracking people at home. Many people living alone may have issues in their daily routines, both due to physical limitations inherent to age or caused by some disease. These people can benefit from some kind of non-intrusive supervision, so the use of cameras or regular visits by social workers is not considered as a first option. A computational system that makes a regular tracking of the subject will be useful in these cases. The daily routine can be monitored and any anomaly can be detected. The point is to select which locations would be considered of interest and the system will report in which zones the subject has been stayed and for how long. The system must be cost-effective, easy to manipulate and install by non skilled people.

This paper presents the results of the first stage of this project, specifically a comparison of different algorithms to process RSS information obtained from Bluetooth in order to obtain information about location of a mobile object. The paper is organized as follows: Section 2 contains some information about Bluetooth and its use for indoor location. Section 3 describes the experimental set-up, the protocol followed for data collection and the algorithms we tested. Section 4 includes the experimental results and some discussion about them. Finally, in section 5 some conclusions are given and future improvements are proposed.

2   Related Work

As previously said, different techniques and technologies have been used for indoor location [9, 2]. The reasons for choosing Bluetooth are that is it easy, cheap and it can be embedded in very small devices. Bluetooth was the technology used in many projects [10–13] based on a wide range of algorithms: proximity-based, trilateration, ad-hoc approaches, received signal strength indicator, fingerprinting and combinations of them and even fusion with others technologies [14, 15].

Proximity-based methods are simple to implement and relatively reliable [16]: if a user is contacted by a beacon, then the user is located in an area defined by the Bluetooth range of transmission. In trilateration, distance from the mobile node to three fixed nodes whose positions are already known is estimated using a radio propagation model; accuracy depends on signal and environmental conditions [7]. The idea is to use RSS values as a measurement of signal attenuation. Although there is no deterministic relationship between distance and RSS values, there is a qualitative trend [17, 3], where the relation can vary due to many factors [18].

Fingerprinting-based methods consist of two stages: set-up stage and online stage. During the set-up stage, a mapping is created using RSS values obtained in distinct positions of the room. In the online stage, object position is estimated based on the database obtained in the first stage [19]. RSS can be affected by diffraction, reflection, and scattering in the propagation [2]. There are several fingerprinting location algorithms based on pattern recognition techniques: probabilistic methods, kNN [20], neural networks, support vector machine (SVM) and others [2]. This paper describes an experiment using Bluetooth and fingerprinting for indoor location. Different techniques related with pattern recognition are evaluated to determinate which one gives more accuracy.

3   Methodology

3.1   Experimental Set-Up

Our experimental testbed was an area assembled as a typical living room. It is a 3.53 x 8.09 m rectangular area of 28.55 m^2. Six zones are marked, each one receiving a letter from A to F. Two Bluetooth devices BT1 and BT2 are placed in the middle of the room: they are equidistant from the borders of the room, and at 70 cm from the floor. This set-up is shown in Figure 1. Both Bluetooth devices are Conceptronic Bluetooth 2.1 USB Nano Adapter with a range of 200 m, class 2 with output power less than 100 mW. Each Adapter is connected to a PC, one running Linux and the other Windows XP. The mobile host is carried by the experimenter, it is a Netbook running Linux with the latest BlueZ protocol stack, and equipped with an USB adapter of the same characteristics of the other two BT devices.

In our experiments, the mobile host is running bluetooth inquiries without connection while standing at different positions, both inside zones A to F and in other positions outside this six zones. RSSI values obtained in the mobile host from BT1 and BT2 are stored. Records contain the MAC of the inquired BT device in range and the associated RSSI value. Inquiries without connection work as a inquiry with connection but they also return the RSSI value at the time of the inquiry. This mode has two important advantages over the connected inquire mode: (1) it does not require an active connection, and (2) RSSI values are free from side-effects caused by power control thus being more reliable and informative [21]. Implementation was done with the PyBluez libraries, that provide access to system's Bluetooth resources and a fine control of the local Bluetooth adapter [22].

RSSI values thus obtained are quite variable even when the mobile host is stationary. Figure 2a shows the probability density estimation of BT1 RSSI values measured in zones A to F. These figures shows that even it is possible to detect some peaks corresponding to RSSI values more repeated, there is too much variability to easily assign RSSI values to distances or positions, as it is also concluded in [12, 18]. In order to overcome this issue and achieve higher accuracy, three pattern recognition algorithms were tested.

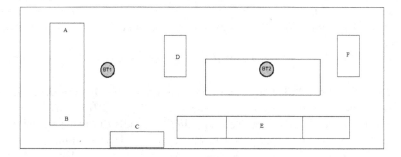

Fig. 1. Experimental testbed

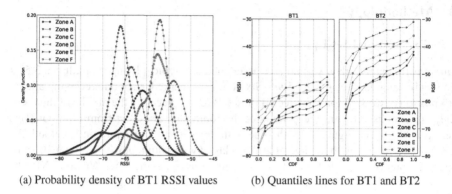

(a) Probability density of BT1 RSSI values   (b) Quantiles lines for BT1 and BT2

Fig. 2. RSSI values

3.2   Algorithms

Three algorithms were compared: the first one, called Cumulative Distribution Function (CDF) uses RSSI values directly. In the other two a wide set of parameters that characterize the RSSI response were calculated as input to a Support Vector Machine (SVM) classifier and a Linear Discriminant Analyzer (LDA).

CDF Comparison. CDF uses quantiles: points taken at regular intervals from the cumulative distribution function of RSSI values. For each of the six reference zones (A-F), 10-quantiles (also called deciles) were calculated from a random sample of RSSI values for each Bluetooth adapter (BT1, BT2). Each zone is then characterized by two vectors $\overline{BT_1^X}$ and $\overline{BT_2^X}$, where X identifies the reference zone. The distribution of these deciles can be seen in Figure 2b. To determinate if a user position corresponds to a reference zone, two vectors $\overline{BT_1}$ and $\overline{BT_2}$ were calculate using randomly selected RSSI from (BT1, BT2). These vectors characterize the position of the user.

The algorithm consists of two steps. First, we calculate for each reference zone X the Euclidean distances of decile vectors:

$$\left(d_1^X, d_2^X\right) = \left(\parallel \overline{BT_1^X} - \overline{BT_1} \parallel, \parallel \overline{BT_2^X} - \overline{BT_2} \parallel\right)$$

So, $\left(d_1^X, d_2^X\right)$ represents the distance between the user and zone X. The lower the distance, the closer to a zone X is the user. If both components of the distance vector are minimal for the same zone, this is recognized as a reference zone. That is, zone X will be identified if

$$X = \underset{x\in\{A,...,F\}}{arg\ min}\ (d_1^x) = \underset{x\in\{A,...,F\}}{arg\ min}\ (d_2^x)$$

If this condition is not fulfilled, a second step is required. First, the nearest BT adapter to the user's position is identified using the medians of vectors $\overline{BT_1}$ and $\overline{BT_2}$ (the higher the median, the closer the adapter). Then, a polling procedure is used, where votes from the nearer BT adapter are given greater scores. If $\overline{BT_1}$ is closer then the closest zone indicated by d_1^X receive a greater score than the closest zone indicated by d_2^X. With this procedure, values d_1^X from both adapters are taken into account, and more powerful signals are given more weight in the decision. This scheme offers the advantage of being easily extensible if using more than two BT adapters.

Advanced Classifiers: SVMs and LDA. With the aim of increasing the complexity of the decision process, the next step was to implement more complex classifiers capable of extracting information from the data in an unsupervised way. In order to select which parameters were going to be used by the classification systems, we calculated a wide set of statistical parameters, including variance, skewness, kurtosis, quartile range, etc. Statistical significance of these parameters was estimated using an Random Forest algorithm [23] and the set was reduced to three parameters: mean, variance and upper quartile. Vectors composed by statistical parameters obtained from RSSI values of both BT adapters were used. As said, we tried two families of classifiers:

- Support Vector Machines (SVM) [24] are a group of supervised learning algorithms which depend on data only through dot-products. In this case, dot-products can be replaced by kernel functions in feature spaces with higher dimensionalities. This methods offer the advantage of being able to generate non-linear decision boundaries using methods designed for linear classifiers [25].
- Linear Discriminant Analysis (LDA) [26] is a statistical technique which tries to classify objects into mutually exclusive and exhaustive groups based on a set of measurable object's features. This is achieved by finding a linear combination of features characteristic from a group of classes.

4   Experimental Results

The total number of readings is 1225. The number of readings per zone is: Zone A 237, Zone B 211, Zone C 193, Zone C 181, Zone D 203 and Zone F 200. This set was divided into four subsets for performing a 4-fold validation, taking 3/4 of the dataset for training and the other 1/4 for testing, and repeating this process four times. For CDF, 30 RSSI values per zone were extracted from the training set in order to calculate $\overline{BT_1^X}$ and $\overline{BT_2^X}$ vectors. For SVM and LDA, training set was used to calculate the statistical parameters which build the features vectors. We used 7 features vectors of 15 RSSI values selected randomly. As there are two adapters BT1 and BT2, the parameters are joined to characterize the zones.

For testing, 15 RSSI values were randomly selected. For CDF, these values were used to obtain $\overline{BT_1}$ and $\overline{BT_2}$, while for SVMs and LDA they were used to obtain the statistical parameters.

First, a test without out-of zone data was performed. Three vectors from each zone were selected, and since there are six zones in a 4-fold validation process, a total of 72 test vectors were used.

Table 1 shows that the three algorithms are capable to discriminate the six zones chosen as reference zones. The best accuracy was achieved using LDA (91.66%). CDF and SVM also achieved good rates: 90.27% and 88.88% respectively. In the SVMs approach, using generic functions to tune hyper-parameters of SVM helped to determine the best hyper-parameters. The best results were obtained using a radial kernel with parameters gamma=0.5 and cost=4.

Table 1. Results without out-of-zone positions

Algorithm	Correct identification	Incorrect identification	Accuracy
CDF	65	7	90.27%
SVMs	64	8	88.88%
LDA	66	6	91.66%

When including out-of-zone data in the experiment, thresholds are required for all algorithms. Setting values for thresholds is the result of a trade-off between sensitivity and specificity. We decided to estimate global thresholds (shared by zones A-F) by averaging the outcomes obtained in the first experiment, without out-of-zone data.

The 4-fold validation process is repeated, with two vectors per zone, giving 48 test vectors. Another 48 out-of-zone vectors are added to the experimental set-up, giving a total of 96 test cases.

Table 2. Results with out-of-zone positions

Algorithm	Correctly identified	Correctly discarded	Incorrectly identified	Incorrectly discarded	Accuracy
CDF	40	44	8	4	87.5%
SVMs	41	44	6	5	88.54%
LDA	43	8	41	4	53.12%

As can be seen in table 2, this procedure achieves good accuracy in CDF and SVM but not in LDA. The CDF approach yielded 87.5% of correct classification, while SVMs achieved 88.54%. LDA gave the lowest rate of success: only 53.12% of correct classification. The percentage fall in the LDA is due to its inability to distinguish among reference zones and nearby positions. Symmetric positions (those which are at similar distances from BT1 and BT2) and out-of-zone positions very close to a zone are the most problematic cases. On the other hand SVM and CDF show more generalization capability, which results in a better performance when dealing with out-of-zone positions.

5   Conclusions and Future Work

This paper describes the results of an experiment conducted to determine if it is possible to track the movements of a person within a room, using only information from Bluetooth transmitters and receivers. We tested three different algorithms and we found that the best is the one based on SVM which achieves a 88.54% of correct classification rate, which was rendered as enough for our purposes.

Our future work will focus on two aspects. The first one is to look for more suitable devices for a real system deployment, that is, small and autonomous Bluetooth transmitters and mobile phones as receivers. The second is to develop a high level layer to process location information with the goal of obtaining higher level semantic information such as usual trajectories or areas where the person stay longer, etc. This layer will allow us to build a behaviour-aware system capable of raising abnormality alarms when variations on the daily routine are detected.

References

1. Ni, L., Yunhao, L., Yiu Cho, L., Patil, A.: Landmarc: indoor location sensing using active rfid. In: First IEEE Int. Conf. on Pervasive Computing and Communications (PerCom 2003), pp. 407–415 (2003)
2. Liu, H., Darabi, H., Banerjee, P., Liu, J.: Survey of wireless indoor positioning techniques and systems. IEEE Transactions on Systems, Man, and Cybernetics, Part C: Applications and ReviewsFhay 37(6), 1067–1080 (2007)
3. Hay, S., Harle, R.: Bluetooth tracking without discoverability. In: Choudhury, T., Quigley, A., Strang, T., Suginuma, K. (eds.) LoCA 2009. LNCS, vol. 5561, pp. 120–137. Springer, Heidelberg (2009)
4. Want, R., Hopper, A., Falcão, V., Gibbons, J.: The active badge location system. ACM Trans. Inf. Syst. 10(1), 91–102 (1992)
5. Priyantha, N., Chakraborty, A., Balakrishnan, H.: The cricket location-support system. In: 6th Annual Int. Conf. on Mobile Computing and Networking. MobiCom 2000, pp. 32–43 (2000)
6. Hossain, A., S.W.S.: A comprehensive study of bluetooth signal parameters for localization. In: IEEE 18th International Symposium on Personal, Indoor and Mobile Radio Communications, PIMRC 2007, pp. 1–5 (2007)
7. Subhan, F., Hasbullah, H., Rozyyev, A., Bakhsh, S.: Indoor positioning in bluetooth networks using fingerprinting and lateration approach. In: International Conference on Information Science and Applications (ICISA), pp. 1–9 (2011)
8. Seco, F., Jimenez, A., Prieto, C., Roa, J., Koutsou, K.: A survey of mathematical methods for indoor localization. In: IEEE International Symposium on Intelligent Signal Processing, WISP 2009, pp. 9–14 (2009)
9. Hightower, J., Borriello, G.: Location systems for ubiquitous computing. Computer 34(8), 57–66 (2001)
10. Anastasi, G., Bandelloni, R., Conti, M., Delmastro, F., Gregori, E., Mainetto, G.: Experimenting an indoor bluetooth-based positioning service. In: 23rd International Conference on Distributed Computing Systems Workshops, pp. 480–483 (2003)
11. Bargh, M.S., de Groote, R.: Indoor localization based on response rate of bluetooth inquiries. In: First ACM International Workshop on Mobile Entity Localization and Tracking in GPS-Less Environments, pp. 49–54 (2008)
12. Forno, F., Malnati, G., Portelli, G.: Design and implementation of a bluetooth ad hoc network for indoor positioning. IEE Proceedings Software 152(5), 223–228 (2005)
13. Jevring, M., de Groote, R., Hesselman, C.: Dynamic optimization of bluetooth networks for indoor localization. In: 5th Int. Conf. on Soft Computing as Transdisciplinary Science and Technology, CSTST 2008, pp. 663–668 (2008)
14. Aparicio, S., Pérez, J., Bernardos, A., Casar, J.: A fusion method based on bluetooth and wlan technologies for indoor location. In: IEEE Int. Conf. on Multisensor Fusion and Integration for Intelligent Systems, MFI 2008, pp. 487–491 (2008)
15. Wang, R., Zhao, F., Luo, H., Lu, B., Lu, T.: Fusion of wi-fi and bluetooth for indoor localization. In: 1st Int. Workshop on Mobile Location-Based Service, MLBS 2011, pp. 63–66 (2011)

16. Elnahrawy, E., Xiaoyan, L., Martin, R.: The limits of localization using signal strength: a comparative study. In: First Annual IEEE Communications Society Conf. on Sensor and Ad Hoc Communications and Networks, pp. 406–414 (2004)

17. Chawathe, S.: Beacon placement for indoor localization using bluetooth. In: 11th International IEEE Conference on Intelligent Transportation Systems, ITSC 2008, pp. 980–985 (2008)

18. Hallberg, J., Nilsson, M., Synnes, K.: Positioning with bluetooth. In: 10th Int. Conf. on Telecommunications, ICT 2003, vol. 2, pp. 954–958 (2003)

19. Pei, L., Chen, R., Liu, J., Tenhunen, T., Kuusniemi, H., Chen, Y.: An inquiry-based bluetooth indoor positioning approach for the finnish pavilion at shanghai world expo 2010. In: 2010 IEEE/ION Position Location and Navigation Symposium (PLANS), pp. 1002–1009 (2010)

20. Orozco-Ochoa, S., Vila-Sobrino, X.A., Rodríguez-Damián, M., Rodríguez-Liñares, L.: Bluetooth-based system for tracking people localization at home. In: Abraham, A., Corchado, J.M., González, S.R., De Paz Santana, J.F. (eds.) International Symposium on Distributed Computing and Artificial Intelligence. AISC, vol. 91, pp. 345–352. Springer, Heidelberg (2011)

21. Raghavan, A., Ananthapadmanaban, H., Sivamurugan, M., Ravindran, B.: Accurate mobile robot localization in indoor environments using bluetooth. In: ICRA, pp. 4391–4396. IEEE (2010)

22. Huang, A., Rudolph, L.: Bluetooth Essentials for Programmers. Cambridge University Press (2007)

23. Breiman, L.: Random forest. Machine Learning 45(1), 5–32 (2001)

24. Cortes, C., Vapnik, V.: Support-vector networks. Machine Learning 20(3), 273–297 (1995)

25. Hsu, C., Chang, C., Lin, C.: A practical guide to support vector classification (2010)

26. Venables, W., Ripley, B.: Modern applied statistics with S. In: Statistics and Computing. Springer (2002)

Dynamically Improving Collective Environments through Mood Induction Procedures

Davide Carneiro, Paulo Novais, Fábio Catalão, José Marques,
André Pimenta, and José Neves

CCTC/DI - Universidade do Minho
Braga, Portugal
{dcarneiro,pjon,jneves}di.uminho.pt,
{pg19832,pg19805,pg20189}@alunos.uminho.pt

Abstract. In our daily living, the environment surrounding us influences us as much or more than we influence it. Whether it is a domestic, leisure or working environment, its conditions will certainly have short and long-term effects on aspects such as stress, mood or fatigue, which will in turn influence indicators such as productivity, quality of work, quality of life, personal/group performance or even health. In this paper a dynamic environment is proposed that, based on the behavioural analysis of its users, will adapt its conditions to improve particular indicators. This will result in better working environments, with an impact on the quality of the work produced.

Keywords: Intelligent Environments, Stress, Fatigue, Mood Induction Procedures.

1   Introduction

The interest in designing harmonious environments for people to live and work in is not new. Some of the earliest evidences of practices aimed at improving physical environments date back to the Neolithic period, which later resulted in popular trends such as Feng Shui. Superstitions aside, Feng Shui can be seen as the process of designing and building spaces that obey to specific rules. When concluded, such spaces are expected to provide specific benefits to their users, particularly in what concerns health and well-being. Recently, several studies have acknowledged the advantages of such practices [1,2].

This paper builds on a similar yet modern view on the shaping of environments. While Feng Shui and related approaches deal with the design and building of new spaces that will remain static afterwards, this paper focus on dynamic and adaptive environments that respond to the user's (implicit or explicit) state.

The aim is the development of workplaces that are aware of their user's state (defined mostly by his stress and fatigue) and continuously adapt the conditions in order to improve the environment. The actions of the environment will have as main objective to affect the mood of the users in order to improve their state. This will result in more harmonious working environments, with effects

A. van Berlo et al. (Eds.): *Ambient Intelligence – Software & Applications*, AISC 219, pp. 33–40.
DOI: 10.1007/978-3-319-00566-9_5   © Springer International Publishing Switzerland 2013

on interpersonal relationships, productivity, quality of work and health. The accomplishment of such technological evolution may lead to a significant socio-economical impact and open the door to a wide range of application domains, including in leisure, learning or domestic environments.

2   Inducing Moods to Alleviate Stress and Fatigue

Stress is a widely studied phenomenon. Despite this, it continues to be a complex topic, mainly due to its subjective and multi-modal nature [7]. Particularly, this paper focuses on stress in the workplace and its socio-economic impact.

The loss of productivity due to stress can be estimated. [9] puts it between $200 and $400 monthly, in workers with depression (often caused by stress or fatigue). Other studies point out to a direct, linear and negative stress-productivity relationship: the greater the stress, the less productive the workforce [8]. Workplace stress has also personal damaging effects. A relationship between burnout and a broad range of negative health symptoms including physical and emotional exhaustion, has been found in several studies. It results in a lack of energy and enthusiasm, feelings of depression, frustration, hopelessness, and a sense of entrapment [10].

Fatigue is also one of the most significant factors influencing performance and mood, being one of the main reasons for Human error as well. The continuous disregard of the effects of fatigue, seen frequently as normal consequences of our busy lifestyle, may end up affecting our health or even our life [3].

Fatigue can be defined as the inability to maintain an expected level of performance in executing a given task, comprising symptoms such as loss of attention, slow reaction or low performance. It also makes the individual moody, less tolerant and more prone to conflict.

Similarly to stress, fatigue is a non-specific symptom. It can be detected from multiple sources including the profile of the user (e.g. age, gender, professional occupation, consumption of alcohol and drugs), performance and precision indicators (e.g. mouse click/movement precision, work throughput) or attention span (e.g. time spent on actual task being performed versus time spent in other non-related tasks).

Inducing specific moods in human beings has been a topic of research of psychology. However, the relationship between cognition and emotions has not been studied in detail until recently. The interest in this relationship lies in the hypothesis of influencing mental processes that, in turn, affect mood. Such mental processes include, but are not limited to, attention, memory or decision-making.

Many different ways to induce mood exist [11], with varying effects on the type of mood induced, its positiveness or negativeness, its duration or its intensity. Some of the most frequent ones include the use of uplifting music, upsetting or relaxing images, critical feedback or storytelling.

The effects of these techniques can act on several spheres of the individual. Experiments have been conducted to induce mood on users in order to increase their creativity and originality [12]. Activating moods (e.g., angry, fearful, happy, elated) leads to more creative fluency and originality than do deactivating moods (e.g., sad, depressed, relaxed, serene). In [13], the satisfaction of elderly in Virtual Environments was improved by inducing joy and relaxation through the use of exercises for generating positive-autobiographic memories, mindfulness and slow breathing rhythms.

3   Acquiring Contextual Features from Behavioural Analysis

The study of stress or fatigue, including their causes and symptoms, has been a topic of disciplines such as Medicine or Psychology. Traditionally, data about users is acquired either through self-reporting mechanisms (generally questionnaires) or through the use of physiological sensors.

The former has known disadvantages, namely: (1) people often lie or exaggerate responses; (2) questionnaires are static (responses of a past questionnaire will not change if the environment changes); (3) are inadequate to represent certain types of complex information (e.g. emotions, behaviours, feelings) and (4) are dependent on the formalization and interpretation of the questions.

The latter is a much more precise approach, having however disadvantages of its own: (1) physiological sensors are generally invasive, with the user physically connected to one or more devices; (2) invasive sensors may have an effect on the results, possibly invalidating them (the sheer consciousness of the individual being measured may influence what is being measured).

The development of an environment that can adjust, in real-time, to significant changes in its context thus calls for non-invasive and dynamic methods of data acquisition. Moreover, given the subjective nature of the aspects under study, such methods should also be personalized.

Thus being, the approach devised focuses on the non-invasive analysis of the users' behavioural patterns. In previous work by the research team, it has been studied how stress affects the way users interact with handheld devices and the way they move inside the environment [5,6]. This approach, which was implemented in the past and will be implemented now, comprises the collection of data on several moments, with different configurations of the independent variables (e.g. stressors, fatigue). The data collected in the different moments is then analysed in search for statistically significant differences that can be the result of the independent variables. It was already demonstrated that stress does influence significantly the way we move and the way we hold and interact with a handheld device [6].

3.1   Features under Study

A similar process is now being implemented in a larger scale, including more devices and features. Besides the features related to handheld devices that were already studied [6] and are out of the scope of this paper, the focus is now on the patterns of interaction with keyboards and mouse. The data collected from these devices results in the following features:

- KEYDOWN TIME - time spent between the key down and key up events;
- ERRORS PER KEY PRESSED - number of times the backspace key is pressed, versus the keys pressed;
- MOUSE VELOCITY - velocity at which the cursor travels;
- MOUSE ACCELERATION - acceleration of the mouse at a given time;
- TIME BETWEEN CLICKS - time spent between each two clicks;
- PRECISION - a measure of the number of clicks on active controls versus the number of clicks on passive areas;
- TOTAL EXCESS OF DISTANCE - excess of distance travelled by the pointer between each two consecutive clicks;
- AVERAGE EXCESS OF DISTANCE - average of the excess distance travelled by the pointer between each two consecutive clicks;
- DOUBLE CLICK SPEED - speed of the double click;
- NUMBER OF DOUBLE CLICKS - number of double clicks in a time frame;
- DISTANCE WHILE CLICKING - distance travelled by the mouse while dragging objects;
- SIGNED SUM OF ANGLES - how much the pointer "turned" left or right during its travel;
- ABSOLUTE SUM OF ANGLES - measures, absolutely, how much the pointer "turned" during its travel;
- SUM OF DISTANCES BETWEEN PATH AND STRAIGHT LINE - between each two consecutive clicks, measures the distance between all the points of the path travelled by the mouse and the closest point in a straight line (that represents the shortest path) between the coordinates of the two clicks;
- AVERAGE DISTANCE BETWEEN PATH AND STRAIGHT LINE - same as above, but provides an average value of the distance to the straight line.

All these features are computed from the data provided by the computers in the environment concerning the events of the keyboard and the mouse. For the purpose, each event has an associated timestamp. Figure 1 depicts the visualization of the data collected for a particular user.

3.2 Architecture

Figure 2 depicts the architecture envisioned to tackle a problem with such constraints. It is composed of six interrelated layers, each one building on the lower-level one and providing services to the upper one. Let us now describe each of them, following a bottom to top.

A working environment, in itself, is composed of users and devices. Users are the key part of the Ambient Intelligence paradigm [4] and are represented computationally by a profile, which includes personal information, preferences, needs, background, context, among other aspects. Users can interact with devices

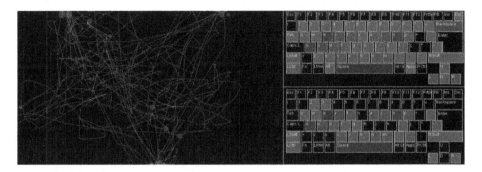

Fig. 1. Visualization of the data collected for each user. The leftmost image depicts the path travelled by the mouse, the areas where the moused clicked more and small vectors depicting velocity and acceleration. The two rightmost images depict the data concerning the keyboard: the topmost depicting the most used keys and the lowermost the keys where there were more typing errors.

and can sense the actions of the devices on the environment. Devices may provide some information about the environment (in which case they become sensors) or may act on the environment (in which case they become actuators). Their type may be very different and their availability may change with time.

The first layer is the *Data Acquisition* one. This layer is responsible for receiving information describing the behavioural patterns of the users. Essentially, it receives data from multiple sources (e.g. computer, handheld device, video camera) and creates the corresponding software objects that can be easily accessed by the superior layers. Each source of data provides the necessary values and the timestamps in which they were measured.

The *Sensor Fusion* layer is responsible for synchronizing the data from the lower layer. Synchronization is performed via timestamps. After this layer, it is possible to know, in a given instant, what was the state of the environment and their components.

The following layer is the *Feature Extraction* one. This layer takes the output from the previous one and generates the features described in subsection 3.1, which already provide some meaning about the environment.

Next, the *Feature Selection* layer takes the features and selects the most suited ones. This process of selection is based on the availability of the sources of the data, the quality of the data and the problem being dealt with.

After the interesting features have been selected, the *Pattern Recognition* layer will compute the highest-level information available on the architecture. This information describes how close the distribution of the data being gathered is to each of the known data models. That is, if the distribution of the data concerning a given user is very similar to the distribution of the data that was collected from that same user when he was stressed or tired, that user is most likely stressed or tired right now.

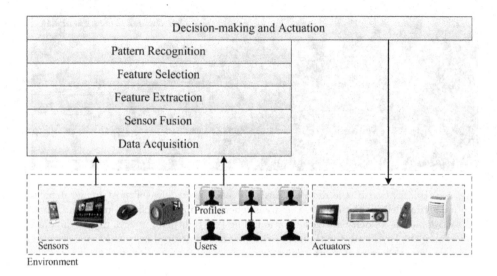

Fig. 2. The layered architecture of the environment. Each layer builds on the lower one and provides functionalities for the upper one.

4   Acting on the Environment

The topmost layer of this architecture encompasses decision-making and influence of the environment. The influence on the environment is achieved through the so-called Mood Induction Procedures (MIPs). The MIPs considered can act at two different levels: (1) at a User Level, including decisions to influence a specific user, with an isolated objective such as calming him down; (2) at a Group Level, including decisions to influence a group of users in its whole (e.g. control of temperature or noise). Given this, MIPs are organized as:

– USER-AREA MIPS
  • AUTOBIOGRAPHICAL RECALL - each user has in his personal space a USB digital photo frame. The contents displayed are controlled by this layer. By looking at pictures of specific objects, people or past personal experiences of the user, memories and emotions linked to those objects, events and people will be experienced by the user, inducing a positive change on user's mood;
  • COLOR SCHEMES - the color scheme can be changed by this layer in order to induce specific states on the user. As an example, dark tones of blue and green are generally recognized to relax people;
  • MUSICAL SELECTION - in the case of the user being allowed to work with headphones, the musical selection may be selected by this layer in order to induce specific feelings on the user;
  • INDIVIDUAL RECOMMENDATIONS - individual recommendations can be issued in the form of notifications that aim to improve the mood of the individual user (e.g. take a break, play a given game for a while).

– GROUP-AREA MIPs

- ENVIRONMENTAL ACTUATORS - this group of devices controls the levels of temperature and luminosity, which are closely related to stress and fatigue. It does it by interacting with air conditioning devices and the lighting system. The control of the electric window blinds can also be used with some purposes (e.g. open them to distract people for a while and increase creativity, close them to make people focus on their tasks);
- ENVIRONMENTAL SOUNDS - the system uses the installed sound system to induce specific moods on the group of users (e.g. agitation, calm) by playing specific types of music or sounds;
- LEVEL OF NOISE - when the level of noise is too loud the system may choose to stop eventual environmental sounds or to show notifications to the users;
- VIDEO PROJECTORS - video projectors are used to display specific colors, shapes, images or videos on the unobstructed white walls of the environment, in order to produce an effect on the global mood of the users;

Using these actuators, the system being devised is able to induce specific moods on the users, with the objective of maximizing some collective goal. This goal can be to achieve a more harmonious and calm environment (traditionally this is desirable for long-term environments) or it can be to induce agitation, creativity or even conflict (such approaches are often desirable for brainstorming sessions, for example).

5 Conclusions and Future Directions

In this paper the negative impact that continued stress or fatigue can have on the workplace was briefly analysed, namely in terms of the productivity, quality of the work, interpersonal relationships and even personal health.

An architecture that is now under implementation was proposed to analyse the interaction patterns of the users in order to search for (personal or collective) signs of stress or fatigue, and then act on the environment to alleviate their effects. Interaction patterns are analysed in real-time, in a non-invasive way, and stress or fatigue level is estimated through the similarity between the pattern being analysed in real-time and known previously classified patterns.

Work is now being done on the topmost layer of the architecture, in order to develop algorithms that can maximize the collective sense of satisfaction of the users, who possibly have conflicting goals, so that the actions on the environment can be optimal. At a user-level, these actions will focus on playing pleasant or relaxing images and/or musics on the personal space, configuring color schemes, among others. The information concerning personal preferences regarding these issues is currently acquired through questionnaires. Work is now also being done to implement profiling techniques that can acquire such information. At a group-level, actions will focus on the improvement of the quality of the environment, concerning aspects such as the noise level, the temperature, the luminosity or even colors or images being displayed in the walls. The rationale to act on the environment will have its foundations mostly on psychology and on occupational ergonomics.

Acknowledgements. This work is funded by National Funds through the FCT - Fundação para a Ciência e a Tecnologia (Portuguese Foundation for Science and Technology) within projects PEst-OE/EEI/UI0752/2011 and PTDC/EEI-SII/1386/2012.

The work of Davide Carneiro is also supported by a doctoral grant by FCT (SFRH/BD/64890/2009).

References

1. Mak, M.Y., Thomas, S.: The art and science of Feng Shui - a study on architects' perception. Building and Environment 40(3), 427–434 (2005)
2. Henderson, P.W., Cote, J.A., Leong, S.M., Schmitt, B.: Building strong brands in Asia: selecting the visual components of image to maximize brand strength. International Journal of Research in Marketing 20(4), 297–313 (2003)
3. Schimdt, R.E., Richter, M., Gendolla, G.H.E., van der Linden, M.: Young poor sleepers mobilize extra effort in an easy memory task: evidence from cardiovascular measures. Journal of Sleep Research 19(3), 487–495 (2010)
4. Aarts, E., Grotenhuis, F.: Ambient Intelligence 2.0: Towards Synergetic Prosperity. Ambient Intelligence and Smart Environments 3(1), 3–11 (2011)
5. Carneiro, D., Carlos Castillo, J., Novais, P., Fernãndez-Caballero, A., Neves, J., López, M.: Stress monitoring in conflict resolution situations. In: Novais, P., Hallenborg, K., Tapia, D.I., Rodríguez, J.M.C. (eds.) Ambient Intelligence - Software and Applications. AISC, vol. 153, pp. 137–144. Springer, Heidelberg (2012)
6. Carneiro, D., Carlos Castillo, J., Novais, P., Fernãndez-Caballero, A., Neves, J.: Multimodal Behavioural Analysis for Non-invasive Stress Detection. Expert Systems with Applications 39(18), 13376–13389 (2012)
7. Selye, H.: The Stress of Life. McGraw-Hill (1978)
8. Spielberger, C.D.: Job Stress Survey. In: The Corsini Encyclopedia of Psychology. John Wiley & Sons (2010)
9. Kessler, R.C., Barber, C., Birnbaum, H.G., Frank, R.G., Greenberg, P.E., Rose, R.M., Simon, G.E., et al.: Depression in the workplace: effects on short-term disability. Health Affairs 18(5), 163–171 (1999)
10. Kagamimori, S., Nasermoaddeli, A., Wang, H.: Psychosocial stressors in inter-human relationships and health at each life stage: A Review. Environmental Health and Preventive Medicine 9, 73–86 (2004)
11. Jallais, C., Gilet, A.: Inducing changes in arousal and valence: Comparison of two mood induction procedures. Behaviour Research Methods 42(1), 318–325 (2010)
12. De Dreu, C.K.W., Baas, M., Nijstad, B.A.: Hedonic tone and activation level in the mood-creativity link: toward a dual pathway to creativity model. Journal of personality and social psychology 94(5), 739–756 (2008)
13. Baños, R.M., Etchemendy, E., Castilla, D., García-Palacios, A., Quero, S., Botella, C.: Positive mood induction procedures for virtual environments designed for elderly people. Interacting with Computers 24(3), 131–138 (2012)

CAFCLA: An AmI-Based Framework to Design and Develop Context-Aware Collaborative Learning Activities

Óscar García, Ricardo S. Alonso, Dante I. Tapia, and Juan M. Corchado

Department of Computer Science and Automation,
University of Salamanca, Plaza de la Merced, s/n, 37008, Salamanca, Spain
{oscgar,ralorin,dantetapia,corchado}@usal.es

Abstract. Ambient Intelligence (AmI) promotes the integration of Information and Communication Technologies (ICT) in daily life in order to ease the execution of everyday tasks. In this sense, education becomes a field where AmI can improve the learning process by means of context-aware technologies. However, it is necessary to develop new tools that can be adapted to a wide range of technologies and application scenarios. Here is where Agent Technology can demonstrate its potential. This paper presents CAFCLA, a multi-agent framework that allows developing learning applications based on the pedagogical CSCL (Computer Supported Collaborative Learning) approach and the Ambient Intelligence paradigm. CAFCLA integrates different context-aware technologies, so that learning applications designed, developed and deployed upon it are dynamic, adaptive and easy to use by users such as students and teachers.

Keywords: Ambient Intelligence, Mobile technologies, Computer Supported Collaborative Learning, Context-Aware Learning.

1 Introduction

In recent years there has been a technological explosion that has flooded our society with a wide range of different devices [1]. Moreover, the processing and storage capacity of these devices, their user interfaces or their communication skills are improved day by day. Thanks to these advances, we are currently surrounded by technology that has changed our habits and customs [2]. All this has caused the apparition of new fields such as Ambient Intelligence, whose main objective is to simplify the use of technology to improve people's quality of life [3].

Education is one of the areas in which Ambient Intelligence presents a greater potential as it provides new ways of interaction and communication between individuals and technological systems [4]. The usage of Information and Communication Technologies (ICT) has been present in educational innovations over recent years [4], modernizing the traditional transmission of contents through electronic presentations, email or more complex learning platforms such as Moodle or LAMS and fostering collaboration between students (Collaborative Learning) [5]. Beside the use of those general-purpose tools in education, other tools that make more specific use of

A. van Berlo et al. (Eds.): *Ambient Intelligence – Software & Applications*, AISC 219, pp. 41–48.
DOI: 10.1007/978-3-319-00566-9_6   © Springer International Publishing Switzerland 2013

technology have appeared. This applies to those that make use of Context-awareness information and ubiquitous computing and communication, fundamental parts of Ambient Intelligence [6].

The inclusion of context-awareness in educational scenarios and processes refers to Context-aware Learning [7], a particular area of application of Context-aware Computing [8]. Moreover, the ability to characterize and customize the context that surrounds a learning situation at a certain time and place provides flexibility in the educational process. This way, learning does not only occur in classrooms, but also in a museum, park or any other place [9], obtaining ubiquitous learning spaces. Thus, there is an extensive literature that addresses the problem of this kind of learning, highlighting those works that attempt to solve contextual information acquisition and providing data to users [9]. The use and integration of different technologies and the approach to specific learning activities characterize these solutions. However, the complexity of understanding and use of the technology and solutions in the aforementioned works does not allow a wide use of them.

This paper presents CAFCLA, a framework aimed at designing, developing and deploying AmI-based educational scenarios. Teachers are able to characterize the context where the learning activity will occur through the creation of a world model in which locate data collectors (e.g., sensors), identify and characterize areas of interest (e.g., paintings in a museum), etc. Moreover, the collaboration between students and the customization of the information available is also provided and can be integrated in the activity design. The framework is supported by a multi-agent architecture that provides intelligence to the learning process by helping to manage the activity, all the communications involved, the context-awareness and the collaboration between students and teachers. In addition, developers and technicians benefit from the Application Programming Interface and the formal schemas provided by CAFCLA.

The following section describes the background and problem description related to the presented approach. Then, the main characteristics of the context-aware technologies that CAFCLA integrates are described. Finally, the conclusions and future work are depicted.

2 Background and Problem Description

Providing contextual information and fostering collaboration between students benefit the learning process [1]. Moreover the combination of Collaborative and Context-aware Learning naturally leads to thinking about ubiquitous learning spaces, characterized by "providing intuitive ways for identifying right collaborators, right contents and right services in the right place at the right time based on learners surrounding context such as where and when the learners are (time and space), what the learning resources and services available for the learners, and who are the learning collaborators that match the learners' needs" [11].

A better understanding of environment through technology allows educators to customize the content provided to students. Similarly, technology facilitates the interaction with the environment and between students. This should be reached in a way as transparent and ubiquitous as possible. The technologies used for the collection of contextual information and for the communication between different devices are the cornerstone of the different works presented here. Literature about Context-aware Learning proposals has been deeply reviewed in this work. Some of the most representative works are classified in this paper, following technological criteria related to communications and data collection.

A first approach to provide contextual information is "tagging the context". Even though RFID (Radio Frequency IDentification) is the most spread technology [14], there are other technologies such as NFC (Near Field Communication) or QR Codes (Quick Response Codes) [12] which are growing fast. As can be seen in the usage of Active RFID, both location and context-awareness are closely related: knowing precisely location of objects and people allows determining what is surrounding them and, consequently, characterizing the context in which they are involved. GPS (Global Positioning System) is the most used technology to provide location in Context-aware Learning [13]. This location system provides a high accuracy level and is currently implemented in a wide range of smart phones and mobile devices. In those cases, the mobile device provides a position to the system. However, most of those works do not implement a specific case of use, but propose a general purpose model in which GPS technology is included to facilitate the provision of contextual data.

Furthermore, GPS technology does not work indoors because of the direct vision necessary between satellites and devices. However, indoor environments are very common in learning: museums, laboratories or the school are places where activities that require mobility can be developed. Trying to cover this lack, different location systems based on Active RFID [14] or Wi-Fi [10] are used. Both cases the performance of systems is similar: student's position is determined by the access point which is providing coverage in each moment. This type of approach has significant limitations when developing context-aware learning activities: the location accuracy is too poor. This situation presents an important problem when areas where context information is different are close (e.g., two paintings in a museum).

The review of the literature evidences some lacks in the Context-aware Learning systems proposed until now. Even some works try to combine different technologies to cover as much situations as possible [10], most of them only cover specific learning situations, as those where tagging context with RFID/NFC [12] is necessary or those where learning occurs outdoors [13]. The combination of both situations is only addressed by M2learn [10]. However, this solution does not provide a precise and efficient location systems or the possibility to integrate wireless sensor networks, except for RFID systems.

However, none of the solutions mentioned before takes into account Ambient Intelligence guidelines. The proposed solutions focus their work on the architectural description, framework developers or end-user applications whose designers have not taken into account how complex will be them for educators or students. Some aspects such as designing intuitive and attractive interfaces or abstracting end users from the complexity of technology, issues on which Ambient Intelligence pays special attention, are not taken into account. Thus, if these aspects are excluded from each solution's design process, the final result may be rejected by students and educators. For this reason, the design process must take into account, from the beginning, the opinion of all the interested parties [5], that is, educators, designers and developers. This way is easier to accomplish with Ambient Intelligent issues related to user interfaces and usability of final applications.

Moreover, the works analyzed in this review do not include mechanisms for data or communication management. Ambient Intelligence emphasizes the transparency of technologies for users. In addition, technology is used to ease ordinary tasks or improve activities and the quality of life [6]. In this sense, systems that combine different technologies do not facilitate mechanisms to change between them (e.g., different communication protocols) attending to the needs of a situation. Similarly, data have to be managed in an intelligent and efficient way. Most of the literature reviewed does not include this issue, using only standard data repositories that only consider persistency and consistency [1]. Functionalities like data redundancy to solve network failures help to make the system dynamic and benefit data accessibility with independence of the place and the moment.

Even though it is well known that collaboration benefits the learning process [1], collaboration between students is an issue not considered by many proposals [10]. Including mobile devices and wireless communication protocols in any learning design that requires mobility (as discussed in this paper) is nowadays necessary. Mobile devices easily connect each other so including collaboration between students is an easy task, increasing the variety of activities and improving the learning process.

Furthermore, this work is developed following Ambient Intelligence guidelines, such as personalization of the provided context or transparency and ease of use for teachers and users. Moreover, the inclusion of reasoning mechanisms facilitate the personalization of data provision or the communication management of these kinds of complex systems [2].

3   CAFCLA Context-Aware Technologies Integration

CAFCLA is a framework aimed at designing, developing and deploying AmI-based educational scenarios, focusing on collaborative and context-aware activities. The framework integrates a set of wireless context-aware technologies and communication protocols (e.g., GPS, ZigBee, Wi-Fi, or GPRS/UMTS). Those technologies allow establishing collaborative activities based on Ambient Intelligence among students and teachers. In this sense, communication models vary dynamically depending on the activity; for example, following a client-server model to perform a data query or forming an ad-hoc network to gather contextual information. Thus, the contextual information is always available and may be modified every time. Contextual information is useful in the educational process, facilitating the acquisition of new knowledge and training. The use of contextual information allows a better understanding of the environment surrounding the learning and student in a given time. With this knowledge, the information received by the students can be dynamically optimized, customized and adapted to their needs and requirements. In order to provide contextual information for a wide range of learning scenarios, CAFCLA integrates three different context-aware technologies. More specifically, CAFCLA integrates indoor and outdoor location capabilities and a platform to deploy wireless sensor networks.

As shown in Figure 1, the main purpose of integrating these technologies is to cover the widest range of possible learning situations. Thus, the contextualization of any object or place will be facilitated by an outside location and / or a representative indoor plan of the place or places where the learning activity is taking place. Thus, the

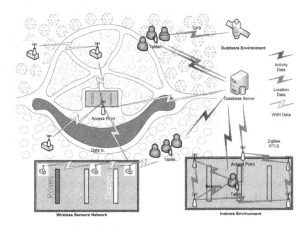

Fig. 1. CAFCLA working schema

system knows when and what information should be provided to students based on knowledge of their positions. Moreover, some contextualization can involve physical measurements, so CAFCLA offers the deployment of wireless sensor networks.

The functionalities of each of the context sensitive technologies integrated into the framework are described in the next sections.

3.1   Wireless Sensor Networks

The integration of a platform to deploy sensor networks in the framework is useful to cover situations in which the contextualization of an environment requires the collection of physical quantities. CAFCLA integrates the n-Core platform [15], which allows the integration of multiple sensors (e.g., temperature, humidity or pressure). The sensors form a mesh network through which data are sent to the access point that sends information to the CAFCLA data server. There, information is stored and processed. Moreover, the sensors can be connected with other ZigBee devices (e.g., a laptop or Tablet PC) and share the data they collect through an ad-hoc connection established in that moment for that purpose. In this case, educators must decide the location of each sensor, the data type and how often data have to be collected and sent to CAFCLA data server. So CAFCLA records where each sensor is placed and implements the protocol to communicate with other sensors, users or data server. Students will be able to receive data from sensors as they approach them by forming an ad-hoc network. Furthermore, the system is aware of which student has approached, so that the information provided can be customized or filtered.

3.2   Outdoor Real Time Locating System

The outdoor location system, specifically the GPS positioning system, is fully integrated into a wide range of mobile devices. For this reason, it is easy to integrate this technology into learning activities. This functionality requires a GPS device and maps platform like Google Maps or OpenStreetMap. CAFCLA integrates all the necessary logical background that may be available on the system and hides all the complexity inherent to the use of this technology to educators. When designing an activity, teachers draw an area on the map. There, all the contextual information related to the area is placed, including different versions of information that are used in different activities or by different users. The system is capable of associating an area to one or more descriptions, so that personalization is easy to achieve. During the development of the activity, students use an integrated GPS device that transmits its position continuously. When the student enters into a characterized area or approaches an object of interest, he or she receives contextual information according to the design of the activity.

3.3   Indoor Real Time Locating System

CAFCLA also provides an indoor Real Time Locating System. The main reason to include this technology is the technical failure of the GPS system to determine the position of users indoors. The Real Time Locating System is based on n-Core Polaris [15], a system that uses the ZigBee wireless communication protocol and that determines the position of users with up to 1 meter accuracy. The n-Core platform facilitates the localization process. The area where the tracking system is deployed is equipped with a set of beacons called n-Core Sirius D. These beacons are able to communicate and send information about the location of a student to the network access point. Each student has a ZigBee device called n-Core Quantum that communicates with the beacons closest to his position. Beacons collect different types of signals sent by mobile devices and send them to the access point. The access point sends all information to the activity server where a location engine calculates the position of the student. Teachers include any kind of information related to any area and students can receive the same way as it is done with the GPS tracking system. Thus, the complexity for educators is reduced and they only have to worry about what information is included in the system regardless of where they perform contextualization.

4   Conclusions and Future Work

The use of Information and Communication Technology in the different areas has increased in recent years thanks to the emergence in society of mobile devices, easy access to currently existing technology and the many features they present, such as communication protocols and context-aware technologies. However, it is difficult to develop applications to squeeze all the potential offered by technology, especially when the main objective is the development of technological applications that are transparent to users, as is suggested by the paradigm of Ambient Intelligence. CAFCLA is a framework designed within the education field with the objective to design and develop a set of tools that provide a basis for designing, developing and implementing Ambient Intelligence based collaborative learning activities that use contextual information. CAFCLA is a framework that integrates different context-aware technologies, such as Real Time Locating Systems, and several communication protocols that abstract educators and developers of context-aware collaborative learning activities from the complexity of the use of different technologies simultaneously. In this case, CAFCLA focuses on provide a set of tools and methods to teachers, developers and technical staff in order to easily design, develop and deploy this type of learning activities. CAFCLA has been designed following the guidelines established by AmI. Requirements such as adaptation, context awareness, anticipation or reasoning have been covered by the implementation of different context-aware technologies that allow the framework to be able to cover a wide range of learning scenarios. Moreover, CAFCLA presents an innovative way to design and develop learning activities, taking into account all staff involved in the process, facilitating the tasks each one is responsible. Future work includes the design, development and deployment of a specific use case where all the features of CAFCLA are implemented. This work will be developed by different teachers and developers in order to compare the results reached by all of them and evaluate the framework in different real scenarios. So that, all the features implemented by CAFCLA will be evaluated and improved thanks to the feedback of final users.

Acknowledgments. This work is supported by the Spanish Ministry of economy and competitiveness.

References

[1] García, Ó., Tapia, D.I., Alonso, R.S., Rodríguez, S., Corchado, J.M.: Ambient intelligence and collaborative e-learning: a new definition model. Journal of Ambient Intelligence and Humanized Computing, 1–9 (2011)

[2] Jorrín-Abellán, I.M., Stake, R.E.: Does Ubiquitous Learning Call for Ubiquitous Forms of Formal Evaluation?: An Evaluand oriented Responsive Evaluation Model. In: Ubiquitous Learning: An International Journal. Common Ground Publisher, Melbourne (2009)

[3] Tapia, D.I., Abraham, A., Corchado, J.M., Alonso, R.S.: Agents and ambient intelligence: case studies. Journal of Ambient Intelligence and Humanized Computing 1(2), 85–93 (2009)

[4] Scardamalia, M., Bereiter, C., McLean, R.S., Swallow, J., Woodruff, E.: Computer-Supported Intentional Learning Environments. Journal of Educational Computing Research 5(1), 51–68 (1989)

[5] Gómez-Sánchez, E., Bote-Lorenzo, M.L., Jorrín-Abellán, I.M., Vega-Gorgojo, G., Asensio-Pérez, J.I., Dimitriadis, Y.A.: Conceptual framework for design, technological support and evaluation of collaborative learning. International Journal of Engineering Education 25(3), 557–568 (2009)

[6] Traynor, D., Xie, E., Curran, K.: Context-Awareness in Ambient Intelligence. International Journal of Ambient Computing and Intelligence 2(1), 13–23 (2010)

[7] Laine, T.H., Joy, M.S.: Survey on Context-Aware Pervasive Learning Environments. International Journal of Interactive Mobile Technologies 3(1), 70–76 (2009)

[8] Dey, A.K.: Understanding and Using Context. Personal and Ubiquitous Computing 5(1), 4–7 (2001)

[9] Bruce, B.C.: Ubiquitous learning, ubiquitous computing, and lived experience. In: Cope, W., Kalantzis, M. (eds.) Ubiquitous Learning, pp. 21–30. University of Illinois Press, Champaign (2008)

[10] Martín, S., Peire, J., Castro, M.: M2Learn: Towards a homogeneous vision of advanced mobile learning development. In: 2010 IEEE Education Engineering (EDUCON), pp. 569–574 (2010)

[11] Hwang, G.-J., Yang, T.-C., Tsai, C.-C., Yang, S.J.H.: A context-aware ubiquitous learning environment for conducting complex science experiments. Computers & Education 53(2), 402–413 (2009)

[12] Tan, Q., Kinshuk, Kuo, Y.-H., Jeng, Y.-L., Wu, P.-H., Huang, Y.-M., Liu, T.-C., et al.: Location-Based Adaptive Mobile Learning Research Framework and Topics. In: International Conference on Computational Science and Engineering, CSE 2009, vol. 1, pp. 140–147 (2009)

[13] Driver, C., Clarke, S.: An application framework for mobile, context-aware trails. Pervasive and Mobile Computing 4(5), 719–736 (2008)

[14] Blöckner, M., Danti, S., Forrai, J., Broll, G., De Luca, A.: Please touch the exhibits!: using NFC-based interaction for exploring a museum. In: Proceedings of the 11th International Conference on Human-Computer Interaction with Mobile Devices and Services, MobileHCI 2009, vol. 71, pp. 1–2 (2009)

[15] Nebusens, n-Core®: A Faster and Easier Way to Create Wireless Sensor Networks. n-Core® (2012), http://www.n-core.info (last access: January 16, 2013)

AndroWI: Collaborative System
for Fuel Saving Using Android Mobile Devices

Víctor Corcoba Magaña and Mario Muñoz Organero

Dpto. de Ingeniería Telemática, Universidad Carlos III de Madrid Leganés, Madrid, Spain
`vcorcoba@it.uc3m.es`

Abstract. This paper implements and validates a system to save fuel based on the collaboration of drivers. The system gets the optimal speed pattern evaluating the driving of nearby drivers. A fuzzy logic system is used to assess drivers and the information about nearby vehicles is obtained through WIFI-Direct. Best driver sends the optimal speed pattern to the other vehicles and the mobile device notifies the user through a vibration pattern or speaker if the user should slow down or speed up.

1   Introduction

One way to reduce fuel consumption and greenhouse emission is to change the driving habits. We should apply a set of rules such as: maintaining a constant speed, driving at high-speed, braking and accelerating smoothly. This is known as eco-driving. Applying these driving rules, we can save between 20 and 25% of fuel [1]-[5]. However, the percentage of fuel depends on the type of vehicle. For example, in hybrid vehicles, we can only save 10% of fuel. One of the problems of eco-driving is that drivers are not motivated to apply the rules eco-driving or forget them [5] [6]. Continuous feedback is needed to improve and maintain the effects achieved by learning [7] [8]. Eco-driving assistant can pro-vide the necessary feedback to adopt an efficient driving style. These systems can be classified into two groups: Evaluation systems and systems based on anticipation. Evaluation systems analyze the driving style and propose improvements to save fuel. Anticipation systems focus their attention on predicting the near future for the vehicle in order to be able to anticipate the driver's behavior to optimize it from the point of view of fuel consumption.

The problem is that anticipation systems need information about nearby vehicles and traffic signals. Currently, although there is a standard (IEEE1609) for vehicular networks (VANET), the vehicle manufacturers do not support them and the infrastructure is not deployed on the roads.

However, in recent years the number of phones have increased exponentially and these have multiple connections (LTE, WiFi, Bluetooth and NFC) to allow connection between them. WiFi connection, particularly on devices that support WiFi-Direct protocol, is suitable for building a vehicular network. In this paper, we propose to build a VANET network using WiFi-Direct to exchange information between vehicles. This information will be used to evaluate the driving style of each driver. The driver will get better score is who provides the reference speed. The other drivers will

A. van Berlo et al. (Eds.): *Ambient Intelligence – Software & Applications,* AISC 219, pp. 49–55.
DOI: 10.1007/978-3-319-00566-9_7   © Springer International Publishing Switzerland 2013

receive on its devices the vehicle speed of the reference driver. Fuzzy logic is employed to evaluate the driving style.

The key to save fuel is to minimize the accelerations/decelerations. Although the optimal speed depends on the vehicle, we have observed in real tests that following the speed of a nearby vehicle that is driving better, we will save fuel. This speed profile is a local optimal solution. However, the driver reduces abrupt accelerations/ decelerations, so it improves the fuel consumption.

Currently, there are on-board computers in cars. These computers show the driver which is the appropriate gear based on the speed for efficient to save fuel, through a screen on the dashboard. However, these systems do not show the optimal speed to minimize decelerations taking into account the current situations of the surroundings. Moreover, such systems are available only in modern vehicles. The solution proposed in this paper can be installed in any vehicle regardless of its age.

2   Collaborative System for Fuel Economy Using WIFI-Direct

In this work, we propose to create a VANET network to exchange vehicle information. This information is used to obtain the best driver from the point of view of fuel consumption. Speed pattern from the best driver is sent to other drivers.

In VANET, using WiFi-Direct protocol, a node will be GROUP OWNER. A GROUP OWNER node is similar to an access point. This node will be whoever receives the information from the rest of the nodes to assess the driving.

The GROUP OWNER node uses fuzzy logic and vehicle's telemetries to determine who the best driver is (node reference). For the driving evaluation, we have considered the last ten minutes and we have used the following variables:

- High Speed: The number of times the speed higher than 100 Km/h allows us to measure the aggressiveness of the driver. Many studies have demonstrated that driving at high speed increases fuel consumption.
- Average Vehicle Speed: It is a parameter that can help us to detect if a driver is novel or aggressive. An unusual vehicle speed can indicate that the driver has bad driving habits.
- Average Acceleration: Acceleration involves increase in energy demand and deceleration involves energy loss. Reduce the number and intensity of accelerations (positives or negatives) allows us to save fuel. Frequent and sudden accelerations mean that driver is not driving at a speed appropriate to the road state.
- RPM: The engine speed is directly related to fuel consumption. If the engine speed is high, driver drives with low gear, and therefore, he is not efficient. On the other hand, if the engine speed is low means that the driver is changing gear conveniently.
- Difference Fuel Consumption: Optimal speed is different in each vehicle since it depends on engine, aerodynamics and vehicle weight. To assess correctly the vehicle speed from the point of view of fuel consumption, we calculate the difference between the retrieved real fuel consumption and the fuel consumption supplied by the manufacturer. If the difference is large, the driver will drive inefficient, and otherwise, the driver will be efficient.

- Time: In many cases, time allows us to know if the vehicle speed is well or not. For example, there will be less traffic at night, and therefore, driver can drive faster than at 19:00 pm (rush hour in Spain).

This information is obtained from vehicle diagnostic port (OBDII). Group Owner set the best node like reference node. Then, the reference node will send its speed pattern to GROUP OWNER. Finally, GROUP OWNER node forwards the speed pattern to other nodes.

This scheme can present a problem. The reference node may be behind the remaining nodes. To avoid this problem, the GROUP OWNER node can only choose as reference node the node that has at least one vehicle behind it.

Another main component of the fuzzy systems are the rules. The rules have a strong influence on system behavior. The number of rules should not be very high because in that case the execution time will increase. On the other hand, if the set of rules are very small, the system response is low quality because it cannot model all the situations that may occur during driving. The rules used in this proposal are:

- IF VehicleSpeed is high and Acceleration is low and Rpm is high THEN efficient
- IF VehicleSpeed is low and Acceleration is Low and Time is rushHour1 or rushHour2 or rushHour2 THEN efficient
- IF VehicleSpeed is high and Acceleration is low and FuelConsumption is Low THEN efficient
- IF VehicleSpeed is high and Acceleration is low and RPM is low and FuelConsumption is Low THEN efficient
- IF VehicleSpeed is low and Acceleration is low and RPM is low THEN efficient
- IF Acceleration is low and fuelConsumption is low and timeSpeedHigh is low THEN efficient
- IF VehicleSpeed is high and Acceleration is High and fuelConsumption is high THEN noEfficient
- IF VehicleSpeed is high and Acceleration is High and Time is rushHour1 or rushHour2 or rushHour2 THEN noEfficient
- IF fuelConsumption is high and timeSpeedHigh is high THEN noEfficient
- IF Acceleration is high and fuelConsumption is high THEN noEfficient

These rules have been obtained observing real data recorded on Spain roads. Out-put from fuzzy system has a value normalized between -1 and 1. If the output is -1, it means that the driver is totally efficient and if output is 1 driver is not efficient.

2.1 User Interface

Mobile devices cause distractions and traffic accident. In [9], we can find a review about driver distractions. Therefore, the user interface is a critical aspect in all eco-driving assistants.

In our case, the assistant has two different outputs: using the speaker or vibration pattern. If we use the speaker, the system notifies the user when he should speed up or slow down. The system only warns when the difference between the recommended speed and current speed is greater than 4 Km/h and vehicle is not stopped.

In the case that the user set vibration notifications, the system uses two types of vibration patterns to alert: a set of short vibrations when driver has to accelerate and a set of long vibrations when driver have to slow down. The notifications are only activated when the difference of speed is higher than 4 km/h and the vehicle is not stopped.

3   Results

3.1   Tests Configuration

Tests were made using three Citroen Xsara Picasso 1.6 HDI with different drivers. This vehicle consumes 6.3 L/100 Km on urban road and 4.1 L/100 Km on highway. It engine has 110 CV and weighs 1313 Kg. The route took place between Leganes and Getafe (Spain). The distance of the route was 8.7 km. This route has urban road and highway. Driving tests were made in October, November and January.

VANET Network was built using three devices: Galaxy Nexus, Nexus 7 and HTC One V. Galaxy Nexus has a dual core processor and 1 GB of RAM. Nexus 7 has a processor with four cores and 1 GB of RAM. Htc One V has a mono core processor and 512 MB of RAM.

The OBDLink OBD Interface Unit from ScanTool.Net [10] was used to get the relevant data from the internal vehicle's CAN bus. The OBDLink OBD Interface Unit contains the STN1110 chip that provides an acceptable sample frequency for the system. In our tests, we obtain ten samples per second.

The execution time of fuzzy logic system was 2 milliseconds in the worst device. Building VANET network time depends on the distance at which the devices are. Maximum coverage using these devices is 120 meters, when the Group Owner node is located in the middle of the network. Otherwise, the minimum coverage is 65 meters. To improve the distance, we could use 3GPP networks in combination with VANET network. In [11] the authors propose a solution to integrate a VANET network with a 3GPP networks. When the nodes of the network are at the limit of coverage, connection establishment time was 35 seconds. The average time for establishing the connection during the tests was 20 seconds.

3.2   Results

Table 1 shows the results of the evaluation of the driving style. The results are employed to determine which vehicle to use as reference. For example, in test 1, the speed profile is obtained from vehicle B (0.80). Table 2 captures the telemetry obtained by vehicles following the optimal speed pattern. We can see that using the collaborative system, drivers reduce fuel consumption when following the optimal speed pattern. In addition, average accelerations and decelerations decreases in the most cases. It also reduces the time that vehicle is travelling at high speed (> 100 Km/h) when this is very long (test 3, driver C).

Table 1. Telemetry used for driving evaluation and results

Test	Driver	Average Speed	Speed >100	Average Acceleration	Average Engine Speed	Fuel Consumption	Result
1	A	44.51 Km/h	60 s	0.54 m/s^2	1514.58 R.P.M	7.43 L/100Km	0.89
1	B	41.11 Km/h	64 s	0.50 m/s^2	1419.71 R.P.M	7.44 L/100Km	0.80
1	C	31.99 Km/h	0 s	0.57 m/s^2	1284.28 R.P.M	8.00 L/100Km	1.00
2	A	43.41 Km/h	62 s	0.55 m/s^2	1479.36 R.P.M	6.59 L/100Km	0.90
2	B	46.72 Km/h	51 s	0.55 m/s^2	1508.67 R.P.M	7.43 L/100Km	1.00
2	C	39.93 Km/h	55 s	0.58 m/s^2	1437.52 R.P.M	7.33 L/100Km	1.00
3	A	42.90 Km/h	55 s	0.52 m/s^2	1466.62 R.P.M	6.97 L/100Km	-0.40
3	B	38.31 Km/h	82 s	0.57 m/s^2	1364.33 R.P.M	6.94 L/100Km	1.00
3	C	40.63 Km/h	128 s	0.64 m/s^2	1432.80 R.P.M	7.06 L/100Km	1.00
4	A	39.30 Km/h	55 s	0.65 m/s^2	1529.56 R.P.M	7.47 L/100Km	1.00
4	B	34.53 Km/h	0 s	0.50 m/s^2	1490.22 R.P.M	6.40 L/100Km	-1.00
4	C	38.30 Km/h	80 s	0.56 m/s^2	1457.88 R.P.M	8.02 L/100Km	1.00

The collaborative system helps the driver to improve driving. If the driver is aggressive, as driver B in test 2, the proposed system achieves the driver to relax the driving. In test 2, driver B drove at high average speed (46.72 Km/h) compared to the average speed of other vehicles in the network. But when the driver follows the optimal driving pattern reduces the average speed (42.53 Km/h) and RPM, saving fuel.

On the other hand, if the driver drives is unsafe (driving at low speed and slows down constantly), the proposed system helps to improve the driving and gives it confidence. Driver C in test 1, before activating the system, he is driving at a speed too low (31.89 Km/h) and average acceleration (positive and negative) is high (0.57 m/s2). However, when the driver uses the optimal speed pattern, the average speed is normalized (40.90 Km/h) and average acceleration decreases (0.52 m/s2).

However, we can observe in test 4 that sometimes, when driver try to follow the optimal speed pattern, he drives slows down and accelerates more times than without using the assistant (Driver B and C). However, in such cases, the fuel consumption decreases despite the increase in acceleration.

In order to validate the system, we have done 10 tests. Each tests involved 3 different drivers. Drivers completed the route between Leganes and Getafe twice. In the first time, drivers drove freely without assistant. In the second time, the assistant was activated using speed profile from the best driver of the first time. Drivers were able to save 6.7% of fuel on average and sudden accelerations (higher than 1.5 m/s^2) were down 30.5%. We observed which the major difference introduced by the use of assistant is appreciated when the driver has an aggressive driving style.

Table 2. Telemetry obtained by vehicles following the optimal speed pattern

Test	Driver	Average Speed	Speed>100	Average Acceleration	Average Engine Speed	Fuel Consumption
1	A	41,28 Km/h	89,00 s	0,59 m/s^2	1520,70 R.P.M	7,19 L/100Km
1	B	40,48 Km/h	48,00 s	0,50 m/s^2	1456,21 R.P.M	6,81 L/100Km
1	C	40,90 Km/h	55,00 s	0,52 m/s^2	1466,62 R.P.M	6,97 L/100Km
2	A	43,22 Km/h	76,00 s	0,54 m/s^2	1561,92 R.P.M	6,01 L/100Km
2	B	42,54 Km/h	67,00 s	0,55 m/s^2	1466,05 R.P.M	7,07 L/100Km
2	C	40,55 Km/h	76,00 s	0,58 m/s^2	1464,32 R.P.M	7,20 L/100Km
3	A	41,34 Km/h	51,00 s	0,50 m/s^2	1456,89 R.P.M	6,80 L/100Km
3	B	40,11 Km/h	50,00 s	0,55 m/s^2	1451,62 R.P.M	6,84 L/100Km
3	C	41,20 Km/h	72,00 s	0,55 m/s^2	1456,21 R.P.M	6,81 L/100Km
4	A	39,93 Km/h	55,00 s	0,58 m/s^2	1437,52 R.P.M	7,33 L/100Km
4	B	39,08 Km/h	20,00 s	0,56 m/s^2	1449,04 R.P.M	6,96 L/100Km
4	C	39,71 Km/h	69,00 s	0,60 m/s^2	1452,41 R.P.M	7,06 L/100Km

4 Conclusion and Future Work

Anticipation is the key to save fuel. Avoid unnecessary energy demand allows you to save a lot of fuel and reduce the emission of greenhouse gases. For this purpose, it is necessary to have knowledge about the environment and vehicles that precede us. In this paper, we proposed a collaborative system to improve the driving from the point of view of fuel consumption. The proposed system lets the driver know in advance the appropriate speed for fuel saving. However, the percentage of fuel saving depends on the skill of the driver.

As future work, we aim to evaluate the system in other environments and with other vehicles. In addition, we want to research the effect that has the collaborative system in the safety, because the speed is one of the factors that most influence in traffic accidents and their severity.

Acknowledgments. The research leading to these results has received funding from the ARTEMISA project TIN2009-14378-C02-02 within the Spanish "Plan Nacional de I+D+I", from the European Union's Seventh Framework Programme managed by REA-Research Executive Agen-cy (FP7/2007-2013) under grant agreement n° 286533 and from the Spanish funded HAUS IPT-2011-1049-430000 project.

References

1. "Traffic: civilization or barbarism". The risk Observatory. Institute for security studies (IDES) (2006), http://www.seguretat.org

2. Barbé, J., Boy, G.: On-board system design to optimize energy management. In: Proceedings of the European Annual Conference on Human Decision-Making and Manual Control (EAM 2006), Valenciennes, France, September 27-29 (2006)

3. Van Mierlo, J., Maggeto, G., Van Burgwal, E., Gense, R.: Driving style and traffic measures-influence on vehicle emissions and fuel consumption. Journal of Automobile Engineering (2004)

4. Koskinen, O.H.: Improving vehicle fuel economy and reducing emissions by driving technique. In: Proceedings of the 15th ITS World Congress, New York, November 15-20 (2008)

5. Johansson, H., Gustafsson, P., Hneke, M., Rosengren, M.: Impact of EcoDriving on emissions. In: International Scientific Symposium on Transport and Air Pollution, Avignon (2003)

6. Wahlberg, A.: Long-term effects of training in economical driving: fuel consumption, accidents, driver acceleration behaviour and technical feedback. International Journal of Industrial Ergonomics, 333–343 (2007)

7. Walker, G.H., Stanton, N.A., Young, M.S.: Hierarchical Task Analysis of Driving: A New Research Tool. In: de Proceedings of the Annual Conference of the Ergonomics Society, London (2001)

8. Tulusan, J., Soi, L., Paefgen, J., Brogle, M., Staake, T.: Eco-efficient feedback technologies: Which eco-feedback types prefer drivers most? In: 2011 IEEE International Symposium on World of Wireless, Mobile and Multimedia Networks (WoWMoM), June 20-24, pp. 1–8 (2011)

9. Young, K., Regan, M.: Driver distraction: A review of the literature. In: Faulks, I.J., Regan, M., Stevenson, M., Brown, J., Porter, A., Irwin, J.D. (eds.) Distracted Driving, pp. 379–405. Australasian College of Road Safety, Sydney (2007)

10. OBD2 Adapter, http://www.scantool.net/ (last access: November 26, 2012)

11. Benslimane, A., Taleb, T., Sivaraj, R.: Dynamic Clustering-Based Adaptive Mobile Gateway Management in Integrated VANET — 3G Heterogeneous Wireless Networks. IEEE Journal on Selected Areas in Communications 29(3), 559–570 (2011), doi:10.1109/JSAC.2011.110306

A Query Expansion Approach
Using the Context of the Search

Djalila Boughareb and Nadir Farah

LabGED Laboratory, Computer Science Department
Badji Mokhtar-Annaba University, P.O. Box 12, 23000 Annaba, Algeria
{boughareb,farah}@labged.net

Abstract. In this paper; we propose a solution to one of the most known problems in information retrieval field which is the ambiguity of short queries. In fact, short queries are often ambiguous and their execution by search tools engenders a lot of noise. The proposed contribution consists of a query expansion approach that exploits the recent browsing history of the user and the time parameter to expand short queries based on the feedback returned by the users having search behaviours similar to that of the current user.

Keywords: Information Retrieval, search context, query expansion, recent interest.

1   Introduction

One of the main reasons of query ambiguity is there shortness. Indeed queries of single or two words are often ambiguous and they may refer to more than one domain of interest. As well, the user expresses his information need with a long query, including keywords that suggest finding together in the returned results as the search system can target his need and returns relevant contents. The single word and the two word queries constitute according to the American audience measurement society One stat.com, more than 50% of the whole number of search engine queries.

If we take an example of the single word query "pascal" and the results of their execution on Google[1], We can find seven different topics in the first 10 documents retrieved. The resulted contents are about the mathematician Blaise Pascal, the university that bears his name, the Pascal programming language, a reference index data base and a lot of other different topics. The asked question is from all these different topics which one the user seeks about through the submitted query?

In order to resolve the shortness problem, the query expansion has been proposed to support the user in his search task through adding search keywords to the query in order to disambiguate it and to increase the number of relevant documents retrieved. The relevance feedback and previous similar queries in addition to relevant visited pages have been proposed and treated widely [4,5,7,9,12,13]. In this paper, the choice of expansion terms will depend on the whole recent navigation activity and we try to

---

[1] http://www.google.com/

A. van Berlo et al. (Eds.): *Ambient Intelligence – Software & Applications*, AISC 219, pp. 57–63.
DOI: 10.1007/978-3-319-00566-9_8   © Springer International Publishing Switzerland 2013

suggest terms that are contextually co-occurred with the query terms having the same search context based on the idea that the set of terms that co-occur often in similar query sessions and having similar search contexts may be useful to expand the user query.

The paper will be structured as follows. In section 2, we present the proposed approach. We define in section 3, the notion of query session as well as its terminological representation. In section 4 and 5, we detail the different steps of the approach. In section 6, we present the evaluation of the work and we discuss the obtained results. In the last section we conclude the paper.

2   Proposed Approach

The works in the contextual information retrieval field that attempt to provide satisfying and rich contents are numerous [1,4,5,7,9,10,12,13,14], a detailed state of the art about the different query expansion strategies can be found in [2]. The major difference between our work and the other works is that we attempt to treat this topic according to new aspects: (1) On the one hand, we emphasize to use the query session including a single search query at a time that aims at filling a single information need as the basic source for expansion. (2) The documents visited a few times ago before the query submission called in this paper the recent interests of the user has been exploited as a contextual dimension in combination with the time dimension with a view of identifying similar query sessions. (3) The time dimension has been studied in several works and used in different manners [1,9,14]. Here, and by employing the time dimension we attempt to invest the periodicity criterion that can arise in a lot of queries. (4) On the other hand, past similar query sessions represented a useful source to our query expansion approach, which aims to resolve the short query problem and proposes to expand the user query based on the users' feedback of past similar previous query sessions having similar search context. Beside the relevance feedback, the pages visited a few times ago before the query submission is also used in identifying related query sessions. In fact, we consider similar, all queries having issued in similar contexts and having similar relevance feedbacks.

3   Query Session

We use the notion of a session and a query session differently in our work. A session is a sequence of queries issued by a single user within a small range of time, while a query session includes a single search query and aims at filling a single information need. As our goal is the improvement of the search on the Web, we were interested in sessions embedded at least one search query. We define formally the query session as follows:

$$\forall S \in \left[t, t'\right] \exists QS \in \left[t, t''\right] / t'' \leq t', QS = \left\{RI, q, t_1, FB\right\} / t_1 \in \left[t, t''\right]$$

The query session QS embeds a single query q issued at the time t_1, the whole navigation activity of the user conducted before submitting the query which includes a sequence of pages visited between $\left[t, t_1 - 1\right]$ with the number of clicks C and the visit time of each one T_V, what is called the recent interest of the user $RI = \{(p, C, T_v)_i / i = 1..n, (p, C, T_V)_I \in [t, t_I - 1]\}$ and a sequence of relevant pages retrieved by the query and visited by the user who issued the query, which represent the user's relevance feedback:

$$FB = \{(p, C, T_v)_j / j = 1..m, (p, C, T_v)_j \in [t_I, t'']\}$$

Each query session QS is represented through a terminological representation QS_t that is described below.

3.1   Terminological Representation

This step allows representing each query session as a list of terms extracted from the relevant pages visited within the query session using Treetagger that is one of the most commonly used tools in the natural language processing field. The first 10 terms having high frequency scores are kept. The main terms of the relevant pages visited a few times before the query are ranged in the set RI_t and those belonging to the pages visited a few times after the query submission are ranged in the set FB_t. The feedback of the query q_i includes the pages visited during the time between the submission of q_i and the query q_j submitted immediately after. To ensure that the page is part of the feedback of the query q_i, we use the function $topic(x)$ to check if the page includes the query terms. The terminological representation of a QS is obtained from the union of two sets of terms RI_t, FB_t and the query terms q_t. The function $topic(x)$ returns the domain may be targeted by a query or the topic of a web page referring to the domain descriptors that are 6 vectors containing the most common words in the treated domains. The function $is\text{-}relevant(x)$ calculates the relevance of a given page using the equation (1). After that, the set of pages visited in QS is ranked by relevance and only the two thirds first pages are kept.

$$R(p, QS) = \frac{T_V.C}{T.C_T} \tag{1}$$

The relevance of a given page which has been visited in a given query session is $R(p, QS)$, here T_v measures the visit time of the page p, C measures the number of clicks inside this page and C_T refers to the total number of clicks on all pages visited during the corresponding query session and T is the duration of the query session.

4   Query Session Clustering

In order to realize a query sessions clustering, we gathered similar query sessions into separated sets $SimQS$ using the cosine similarity measure given by the equation (2) [11,13] which is broadly used in information retrieval field.

$$Sim(x, y) = \frac{x.y}{\|x\|.\|y\|} \tag{2}$$

This therefore allows to avoid obtaining a greater number of clusters produced using an unsupervised learning strategy without making the regrouping. Let S a navigation session including at least one search query; we divided S into k query sessions including each one, one single search query $S = \langle\langle RI_1, q_1, t_1, FB_1\rangle,...\langle RI_k, q_k, t_k, FB_k\rangle\rangle$, where $(k \geq 1)$ is the number of queries in the whole session; after that assembled the query sessions by similarity into $SimQS_i$ using the cosine function which resulted 18 different $SimQS_i$. Then, we prepared each $SimQS_i$ for clustering using multi-layer perceptrons MLP, such as each $SimQS_i$ corresponds to one cluster.

4.1   The Dataset

We constructed our own data set by creating a plug-in for the Firefox browser in order to record the user's browsing history from eight machines in a net-space made available to an audience of users. The net-space ensured for each user only one hour of navigation per day; the add-on installed on each machine saves all users' browsing activities that are captured from the browser. During the month of November, 2011, we collected 12 MB of query log including 396 sessions of one hour each one, each session includes 4 queries in average.

4.2   Machine Learning

In our experiments, we opted for a supervised learning strategy using the Multilayer Perceptron (MLP) [8] which is an artificial neural network that gave an important accuracy in classification problems [3,6]. The created network has been used for learning users' query sessions grouped through their similarity degrees into separate sets of similar search behaviours. The used data set includes 1245 entries each one corresponds to a single query session that includes a single query; the data set was subdivided into training and test. The training set consists of 830 entries were used for learning, while the test set which contains 415 entries were used to estimate the test error. The neural network configuration consists of 32 entries while the first sequence of 14 values corresponds to the neural inputs and the last sequence represents the desired neural outputs. The edited MLP with two hidden layers of 16, 15 neurons respectively is trained using the back-propagation algorithm.

5   Expansion Terms

The choice of the expansion terms depends on the whole recent browsing activity and try to suggest terms that contextually co-occurred with the query terms having the same search context. Indeed, the set of terms that co-occurred often in similar query sessions are used to expand the user queries as follows. Let $SimQS_i = \{QS_1,...QS_n\}$ a set of similar query sessions; let $\{QS_1,...QS_l\}$ the set of terminological representation of them. Based on the couple $\langle q_t, FB_t \rangle$, we build the corresponding term co-occurrence matrix M_n^i. Let $SimQS_i$ a set of similar query sessions and let n be the number of terms in $SimQS_i$. The co-occurrence matrix denoted M_n^i corresponds to a square matrix of n rows and n columns. For each new query, find in the corresponding term co-occurrence matrix the 3 first terms having a high co-occurrence frequency with the query terms, and suggests them for expanding the user query.

6   Evaluation

In this step, the contribution of the proposed approach is checked referring to the results provided by Google. We used a sample of 10 users who were invited to conduct searches and to provide after their judgments about the returned results. First, the query is submitted to Google search engine and the relevance of the first retrieved documents is determined by the user who submits the query through a simple form in which he indicates the number of relevant pages in the returned results. Except in the case when the Web page contains no information on the topic desired by the user, or it belongs to another domain to that one targeted by the user, it is considered irrelevant. After that; the system executes the different steps of the expansion algorithm cited in section (5). The expanded query is submitted to Google and the users' judgments are offered again. The precision at the top 5, 10 and 15 documents is measured in both cases through the equation (3) and a simple comparison between results is made to evaluate the effectiveness of the expansion approach.

$$P@n = \frac{R_n}{n} \tag{3}$$

In equation (3), R_n represents the number of relevant documents in the top n results where n takes the values 5, 10 and 15. The table (1) presents the judgment of users about the results provided by the system during this period of time.

Table 1. P@n precision before and after query expansion

Precision	P@5	P@10	P@15	Average
Before expansion	0,58	0,55	0,53	0,55
After expansion	0,59	0,57	0,57	0.58

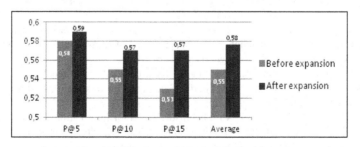

Fig. 1. P@n precision before and after query expansion

The evaluation results of the query expansion approach show that there is not a significant improvement in the relevance of the search results. The main reason is that the terms used in expansion are not always relevant, which is due to the fact that the frequency of co-occurrence of two terms in a given context does not often indicate that each one may be useful for disambiguating the other. Also the frequency degree alone can not be an effective parameter to pick out the relevant terms from the set of co-occurrences.

7   Conclusion

In this paper, we proposed a contextual query expansion approach in order to help the user to better express his information need and furthermore improve the results of the search task on the Web based on the real-time search context. Our study focused on the day and the recent interests of the user as contextual dimensions.

The state in which the user conducts a search on the Web has an effect on its search behaviour, for that reason, the search context must be studied better with the view of discovering new dimensions that can affect the users' search behaviour. There are several areas for future works; first, the context covered in this work may be further extended and combined with other dimensions. We attempt to more refine each group of similar query sessions in order to improve the quality of the co-occurrence relationships between the terms of the same group.

Rreferences

1. Bouidghaghen, O., Tamine, L.: Spatio-Temporal Based Personalization for Mobile search, Engineering, and Intelligent Technologies. In: Next Generation Search Engines: Advanced Models for Information Retrieval, pp. 386–409. IGI Global Publishing, PA (2012)
2. Carpineto, C.: A Survey of Automatic Query Expansion in Information Retrieval. In: Carpineto, C., Romano, G., Bordoni, F.U. (eds.) ACM Computing Surveys, vol. 44(1), Article 1(2012)
3. Ciaramita, M., Murdock, V., Plachouras, V.: Online Learning from Click Data for Sponsored Search. In: Proceedings of the International World Wide Web Conference (IW3C2), WWW 2008, Beijing, China (2008)
4. Fonseca, B.M., Golgher, P.B., Possas, B., Ribeiro-Neto, B.A., Ziviani, N.: Concept-based interactive query expansion. In: CIKM, pp. 696–703 (2005)
5. Kanaan, G., Al-Shalabi, R., Ghwanmeh, S., Bani-Ismail, B.: Interactive and Automatic Query Expansion: A Comparative study with an Application on Arabic. American Journal of Applied Sciences 5(11), 1433–1436 (2008)
6. Klassen, M., Paturi, N.: Web Document Classification by Keywords Using Random Forests. In: Zavoral, F., Yaghob, J., Pichappan, P., El-Qawasmeh, E. (eds.) NDT 2010. CCIS, vol. 88, pp. 256–261. Springer, Heidelberg (2010)
7. Lv, Y., Zhai, C.: Positional relevance model for pseudo-relevance feedback. In: Proceeding of the 33rd International ACM SIGIR Conference on Research and Development in Information Retrieval, SIGIR 2010, pp. 579–586 (2010)
8. Rosenblatt, F.: The Perceptron: A Probabilistic Model for Information Storage and Organization in the Brain. Journal of Psychological Review 65(6), 386–408 (1958)

9. Ruthven, I.: Re-examining the Potential Effectiveness of Interactive Query Expansion. In: Proceedings of ACM SIGIR 2003, Toronto, Canada (2003)
10. Said, A., De Luca, E.W., Albayrak, S.: Inferring Contextual User Profiles–Improving Recommender Performance. In: 3rd RecSys Workshop on Context-Aware Recommender Systems, Chicago, IL, USA (2011)
11. Salton, G., Wong, A., Yang, C.: A vector space model for automatic indexing. Communications of the ACM 18(11), 613–620 (1975)
12. Song, M., Song, I., Hu, X., Allen, R.B.: Integration of association rules and ontologies for semantic query expansion. Journal of Data & Knowledge Engineering 63(1), 63–75 (2007)
13. Wang, H., Liang, Y., Fu, L., Xue, G.-R., Yu, Y.: Efficient query expansion for advertisement search. In: Proceedings of the 32nd Annual International ACM SIGIR Conference on Research and Development in Information Retrieval, pp. 51–58. ACM Press, Boston (2009)
14. Zhao, Q., Hoi, S.C.H., Liu, T.Y., Bhowmick, S.S., Lyu, M.R., Ma, W.Y.: Time-Dependent Semantic Similarity Measure of Queries Using Historical Click-Through Data. In: Proceedings of the 15th International Conference on World Wide Web, pp. 543–552. ACM, Edinburgh (2006)

Guidelines to Design Smartphone Applications for People with Intellectual Disability: A Practical Experience

Raul Igual[1], Inmaculada Plaza[1], Lourdes Martín[1],
Montserrat Corbalan[2], and Carlos Medrano[1]

[1] R&D&I EduQTech Group - Electronics Engineering Department,
Escuela Universitaria Politecnica de Teruel, University of Zaragoza, Teruel, Spain
[2] R&D&I EduQTech Group - Escuela de Ingeniería de Terrassa,
Polytechnic University of Catalonia, Terrasa, Spain
{rigual,inmap,lourdes,ctmedra}@unizar.es,
montserrat.corbalan@upc.edu

Abstract. Applications for smartphones have a great potential to facilitate the lives of people with intellectual disability. In fact, it is possible to design specific applications adapted to their needs. But even in this case, users may experience accessibility issues with some structural elements of smartphones. In this study, we have identified these elements through a 2-month test period with some people with intellectual disability. They used a simple smartphone application that met some needs identified by their caregivers. Through this practical experience, problems with the notification bar and the home, back, menu, search, volume and power buttons have been detected. Potential solutions to overcome these issues have also been proposed.

1   Introduction

People with intellectual disability are characterized by significant limitations both in intellectual functioning and in adaptive behaviour, which covers many everyday social and practical skills [1]. Research has shown that assistive technology can facilitate learning, increase access, and serve as a tool to compensate for specific challenges associated with a disability [2].

In this sense, traditional mobile phones have evolved into modern "smart" phones which combine the communication facilities of cellular phones with the potential of handheld computers [3]. Regarding their use, a recurring theme is the diversity across subjects [4]. Bryen et al. [5] reported that the mobile phone usage rate of adults with intellectual disabilities was much lower than expected. Many mobile phones still lack features that would reduce the level of difficulty and improve access to them. Furthermore, the complexity of some mobile handsets and services can represent barriers to use [6].

Some studies have focused on identifying the features that make the use a complex process. Urturi Breton et al. [7] examined the problems intellectually challenged users might experience when using touch screen mobile phones. They identified issues

A. van Berlo et al. (Eds.): *Ambient Intelligence – Software & Applications*, AISC 219, pp. 65–69.
DOI: 10.1007/978-3-319-00566-9_9   © Springer International Publishing Switzerland 2013

regarding buttons (too small), menus (too many), text size (too small), multi-touch events (too complex) and feedback (not provided). Interfaces should be clear and concise. Other studies also point in the same direction [8][9].

Most authors admit that it is not possible to face people with intellectual disability to smartphones as they were initially conceived [7][8], and we concur with this opinion. However, we cannot renounce using such powerful devices since they have great potential in the field of ambient intelligence. Smartphones, like computers, also run a hardware operating system, which allows for the development of supplemental software applications to be run on the phone [3]. We should take advantage of this potential to make them accessible to people with intellectual disability. It is possible to design specific applications adapted to their needs, with reduced functions, improved usability and enhanced accessibility. But even if an application is adapted, there are structural elements of the smartphones that could hinder its use. These elements are common to all applications and inherent to the mobile devices. In this study, we present a practical experience that has enabled us to identify these elements. The results presented here are general and aim to be of interest to researchers developing smartphone applications for the disabled.

2   Subjects

Three people with intellectual disability were recruited for this study. They were members of the *Agrupacion Turolense de Asociaciones de personas con Discapacidad Intelectual (ATADI)*. The study was approved by the board of the organization and two of their caregivers were involved in the 2-month test period. None of the three participants had used a smartphone before. During the tests, they should interact with a smartphone running an application for people with intellectual disability. Each of the three subjects wore the phone for two months. The tests were performed in the subjects' living environment and they used an application designed according to the needs identified by their caregivers [10] (section 3). Therefore, we did not force interaction with the mobile phones. In fact, this is one of the strengths of this study; we avoided scheduling sessions where users are often asked to perform some predefined actions under the supervision of the researchers. Instead, the subjects handled the phones in their living environment as they considered, without receiving specific instructions. This long real-world evaluation allowed some problems emerge that, in a supervised context, would have rarely been detected. The application with which users interacted is briefly explained in section 3.

3   Methods

The application is intuitive and quite simple. It consists of a button to be pressed in case of emergency. Simultaneously, it locates and stores the position of the phone. In this sense, caregivers can visualize the routes performed by users through their PCs (Figure 1). This last task is accomplished in a transparent way, that is, users' intervention is not required. Their interaction with the phones occurs if they leave some predefined security areas, in which case they are alerted through a message sent to their phones. An alarm tone rings and simultaneously the phone starts vibrating. Users can view the content of the message by switching on the phone screen.

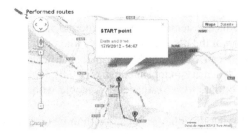

Fig. 1. Screen to visualize the routes performed by the subjects

4   Guidelines to Design Applications for Smartphones

Through this practical experience we could identify which smartphone elements interfere in the use of an application by people with intellectual disability. These elements are represented in Figure 2. This section explains how they affected the operation of the phone during the 2-month test period, describing the actions that can be undertaken.

Fig. 2. Structural elements of a smartphone that could hinder its use

The structural elements of the smartphones that hinder their use are:

- Back, Home and Search Buttons: Users often push these buttons unintentionally. If the back button is pressed, the application is destroyed. When this occurred, none the participants were able to launch it again. In fact, they did not even notice that the application was not running, so they thought it was still active. By contrast, the home button makes the application running on the background and, as in the previous case, users were unable to open it again, so although the location function still worked, the help button was no longer accessible since the application main screen was not displayed.

- Menu Button: By pressing this button a list of commands or facilities is displayed on the screen. It is highly prone to errors, since it requires the interaction with two elements; first, the button itself and second, the option within the menu. None of the 3 subjects involved in the study were able to perform these actions.

- Volume Button: The volume should be controlled internally by the application. Two of the participants accidentally used to switch to vibrate or silent mode which affected the operation of the phone, for example, when receiving a call.
- Notification Area: It can cause confusion since notifications are displayed in a small area (up to 64 density-independent pixels tall). The characteristics of this area make it inappropriate to be managed by people with very low technological skills.
- Power: By long-pressing this button a menu with three options (power off, airplane mode and restart) is displayed. Participants often accessed this menu accidentally due to the need of short-pressing the power button to switch on the screen. None of the three subjects were able to distinguish between a short click and a long click.

To overcome the problems encountered, some design guidelines concerning each one of these elements are provided in table 1. An extra column has been added, reporting on the degree of complexity in carrying out the proposed solutions in an Android device.

Table 1. Solutions proposed to overcome the problems encountered

Structural element	Solution proposed	Degree of complexity
Back button Home button Search button	The normal behaviour of these elements should be overridden, preventing the application from being destroyed (back button), from returning to the main screen (home button) or from accessing to the search option that can cause confusion in the users.	Low. The Android operating system has functions to override the normal operation of these buttons.
Menu button	The menu button should be discarded, and all options included in the main screen of the application. Obviously, this screen should be designed according to the international accessibility standards [11].	None. It is a designer's choice whether to include a menu or not.
Volume button	The volume should be controlled internally by the application. That means, caregivers should decide whether to enable or disable the volume settings, since users may inadvertently change the configuration.	Low. The Android operating system has functions to control the volume settings.
Notification area	The notification area should not be present, filling the whole screen with the content of the application.	None. It is a designer's choice whether to include the notification area or not.
Power	At best, the functionality of the power button should be overridden and a password required for turning off the phone. Since this is a hard task, strategies to prevent the use of the power button are needed. This necessarily leads to the search for alternative ways of switching on the screen. The volume button is a good candidate for this function.	Very high. The power button is a structural hardware element of the phone, and changing its operation, if possible, is a hard task.

Most of the design guidelines given in this paper can be easily implemented. In fact, we have developed a preliminary application with the back, home, search, menu and volume buttons overridden and the notification area removed. As this application lacks of functionality, it should be seen as a framework through which any type of adapted application could be implemented. Thereby, future applications should be contained inside this "framework" application in such a way that if they are designed following the accessibility standards, the whole system will be usable, not being compromised by the structural elements of the smartphones.

5   Conclusion

This study identifies the structural elements of a smartphone that may hinder its use by people with intellectual disability. These elements are the notification bar and the back, home, search, menu, volume and power buttons. When an adapted smartphone application is developed, the functions of these buttons should be overridden to make the system usable. Furthermore, all options within the application should be made accessible from the main screen, designing the interface according to the international accessibility standards [11].

References

1. American Association on Intellectual and Developmental Disabilities (AAIDD), Definition of Intellectual Disability (2012), http://www.aaidd.org (accessed December 2012)
2. Mechling, L.C.: Assistive Technology as a Self-Management Tool for Prompting Students with Intellectual Disabilities to Initiate and Complete Daily Tasks: A Literature Review. Educ. Train. in Dev. Disabil. 42(3), 252–269 (2007)
3. Susick, M.: Application of smartphone technology in the management and treatment of mental illnesses, Master's Thesis, University of Pittsburgh (2011)
4. Falaki, H., Mahajan, R., Kandula, S.: Diversity in Smartphone Usage. In: Proc. of Mobile Systems, Applications, and Services, San Francisco, USA (2010)
5. Bryen, D.N., Carey, A., Friedman, M.: Cell phone use by adults with intellectual disabilities. Intellect. Dev. Disabil. 45(1), 1–9 (2007)
6. Australian Mobile Telecommunications Association (AMTA), Mobile Phone Industry Good Practice Guide: Accessibility for People with Disabilities (2005), http://www.amta.org.au/ (accessed December 2012)
7. de Urturi Breton, Z.S., et al.: Mobile communication for intellectually challenged people: a proposed set of requirements for interface design on touch screen devices. Commun. in Mob. Comput. 1(1) (2012)
8. Verstockt, S., et al.: Assistive smartphone for people with special needs: The Personal Social Assistant. In: Proc. of Human System Interactions, Catania, pp. 331–337 (2009)
9. Lanyi, C.S., et al.: Results of User Interface Evaluation of Serious Games for Students with Intellectual Disability. Acta Polytechnica Hungarica 9(1), 225–245 (2012)
10. Martin, L., Plaza, I., Rubio, M., Igual, R.: Localization Systems for Older People in Rural Areas: A Global Vision. In: Proc. of Distributed Computing and Artificial Intelligence, DCAI, Salamanca, Spain (2012)
11. UNE 13980X Family of standards.: Computer applications for people with disabilities. Computer accessibility requirements (2009)

A Verbal Interaction Measure Using Acoustic Signal Correlation for Dyadic Cooperation Support

Alexander Neumann and Thomas Hermann

Bielefeld University, Universitaetsstrasse 25, 33615 Bielefeld, Germany
`alneuman@cit-ec.uni-bielefeld.de,`
`thermann@techfak.uni-bielefeld.de`

Abstract. We introduce a method for detecting whether two users are engaged in focused interaction using a windowed correlation measure on their acoustic signals, assuming that a continued exchange of verbal turns contributes to anticorrelation of acoustic activity. We tested our method with manually annotated transitions between focused and unfocused interaction stemming from experiments on AR-based cooperation within a research project on alignment in communication. The results show that a high degree and extended duration of speech activity anticorrelation reliably indicates focused interaction, and might thus be a valuable asset for situation-aware technical systems.

Keywords: situation awareness, collaboration, speech activity, data mining, multiscale analysis, correlation.

1 Introduction

Recent developments on technical interactive systems do not only focus on user interfaces that are easy to use but also take the actual usage context into account. Features like ambient light or GPS location information are already used to change the behavior of mobile phones or smart environments, allowing to create situation awareness and adapt the system to the changing environment. Verbal utterances are commonly used either in a rudimentary way to detect general ambient noise or in a complex way which involves speech recognition and semantic parsing. Regarding conversation, using speech only for noise detection ignores its vital role in joint activities [2], while speech recognition often does not fulfill accuracy or speed requirements for reliable information gathering in such a context. It also demands powerful hardware which can involve more than one recording device [7].

We propose a simple and lightweight speech activity correlation approach to reveal verbal dialogue communication patterns which can be used to increase situation awareness for static and mobile cognitive interaction technology. These developments come from the *Augmented Reality based Interception Interface* (ARbInI) which we developed as a system to investigate communication phenomena such as alignment, joint attention and co-orientation in human-human interaction. *ARbInI* was used to collect data in our latest study, a cooperative interaction study where participants had to collaboratively plan fictional building activities and negotiate possible solutions. Besides video and

A. van Berlo et al. (Eds.): *Ambient Intelligence – Software & Applications*, AISC 219, pp. 71–78.
DOI: 10.1007/978-3-319-00566-9_10　　© Springer International Publishing Switzerland 2013

tracking data, the recorded multimodal corpus also includes sound signals from headset microphones that our participants had worn during the experiments.

Exploratory data mining revealed interesting speech activity patterns in these data which we further investigated. Based on 10 dyads from our corpus we developed a correlation measure which depends on noise threshold Θ, silence duration d_p and correlation window size ω, and we tested the algorithm performance against manual annotations of the same data. In the following we will give a brief introduction to our study and the collected data corpus. After that, we will introduce the algorithm and its evaluation and furthermore also show how the algorithm's parameters can be determined from the data. However, we propose that d_p and ω do not have to be adapted to fit varying scenarios.

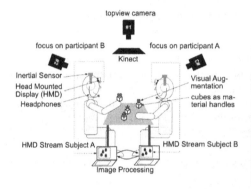

A's field of view   B's field of view

Fig. 1. ARbInI consists of static components such as three DV cameras, a Microsoft Kinect and two to three workstations. Each participant also wears a head-mounted display, a microphone headset and a BRIX motion sensor to measure head movement at high temporal resolution.

Fig. 2. In the ongoing study our participants collaborate to recreate a local lake and its surroundings. *ARbInI* monitors their actions. The markers on top of the wooden cubes are augmented with models representing concepts for possible projects (e.g. *hotel* or *skater park*).

2   Alignment in AR-Based Cooperation

The Collaborative Research Center 673 *Alignment in Communication*[1] investigates the role of alignment and other communication patterns for successful communication. In the subproject C5 *Alignment in AR-based collaboration* we use Augmented Reality (AR) as a technology for communication research which provides new features and methods for this discipline.

Within this context the *Augmented Reality based Interception Interface* (ARbInI) was developed and tested as a monitoring and assistance system in everyday dialogue scenarios [4]. The system allows a direct access to the audiovisual communication channels to monitor and alter information perceived by the users. Combined with other non-verbal communication cues such as gestures, posture and gaze direction these data form a complex multimodal data corpus.

---

[1] www.sfb673.org

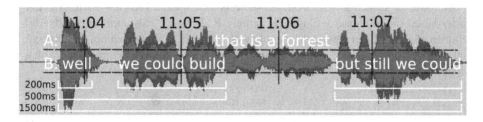

Fig. 3. This is participant B's waveform of a 5 second dialogue sample. Participant A interrupts participant B to deny her suggestion instantly. The orange area of the waveform indicates parts louder than −15 dB which is a sufficient threshold choice here. The speech activity before the interrupt should be merged into a continuous activity but the pause must remain. A short silence duration like 200 ms leaves the activity fragmented; a long one like 1500 ms might close too many gaps.

2.1 ARbInI and Obersee II Scenario

Our system consists of several components which are either positioned around two chairs and a table or worn by the users. All components are shown in Figure 1. The sensors attached to the users contain motion sensing devices from the BRIX toolkit which was developed in our working group [10] and headset microphones to record audio signals. The core component is a video-see-through head-mounted display (HMD) equipped with two Firewire cameras and a display for each eye. Three HD digital video cameras surround the participants, two of them are placed diagonally behind each participant and the third right above the table where also a Microsoft Kinect[2] is located. All data streams can be accessed, stored and manipulated in real-time except for the HD videos which we only record for later analysis.

For the study we have designed a recreation planning scenario which takes place in the surroundings of a lake called Obersee in the city of Bielefeld.

Figure 2 shows the setup from the top with the sketch of the Obersee area in the middle of the table. An important part for our AR approach is the introduction of mediating objects which represent constructions for the participants to use for their planning. They are wooden cubes which are used as "physical handles" with *ARToolkitPlus* [8] markers attached on top. When the system detects a marker it augments the corresponding visual representation of a building or concept on top of the cube as depicted in Figure 2. This feature allows us to monitor, control and manipulate the visual information available to both users separately during the negotiation process at every moment during the experiment [3].

3 Analysis

In the analysis process of the collected data, we investigated speech activity as a feature for measuring the degree of collaboration. We define *speech activity* as any verbal utterance which addresses the speaker's interlocutor with no regards to syntactic or semantic information.

---

[2] www.xbox.com/en-US/kinect

We retrieved speech activity from the subjects' microphone recordings with a *sound finder* based on an audacity plugin by Jeremy R. Brown[3]. This approach reads 100 samples of a signal and detects the sample with the highest volume within this frame k and returns 1 if this sample is louder than Θ. The result was further compressed with a *sample & hold* interpolation to fit the 50 Hz sample rate of our data set.

$$sp[k] = \begin{cases} 1 \text{ if } 10 \cdot \log_{10}(\max(s[i]^2)) > \Theta \\ 0 \text{ else} \end{cases}$$

$$s[i] \in [-1,1], \frac{i}{100} \in [k,k+1], t = \frac{k}{441}\text{sec}$$

(1)

In our case -15 dB has been proven to be a robust and reliable noise threshold which detected all verbal utterances articulated by the speaker without *false positives* like background noise, speech activity of the interlocutor or pure intrapersonal stimulation such as very quiet "hmm" sounds which did not fulfill communication purposes. Certainly, this threshold depends on our special case since used hardware and control parameters (e.g. microphone volume) vary between scenarios.

However, the feature so far leads to fragmented results. For instance even a single word like "friendship" could result in two chunks due to intonation and short pausing between syllables. Therefore, we applied an *erosion* method where gaps within a continued activity are bridged and fragments are merged into a continuous segment if the gap is shorter than a silence duration parameter d_p. We chose this duration with help of the multiscale correlation structure described in section 3.1. Our goal was to ignore small pauses (e.g. "well,... uhm... what about here") but to keep independent statements separated as depicted in Figure 3.

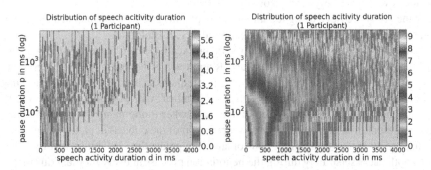

Fig. 4. The color maps show the distribution of speech activity durations on the x-axis and the chosen threshold d_p on the y-axis. The color is the logarithm of the amount of activities with a certain duration. The left plot shows the data of one participant which includes some vertical lines, for instance at 2000, 2500 or 3500 ms of speech activity duration. These lines indicate durations which are very consistent for d_p in range of 250 to 1500 ms. With 20 participants included (as seen on the right) this lines form a "corridor" within this range.

---

[3] audacity.sourceforge.net

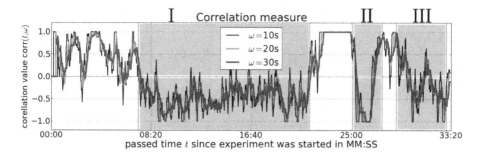

Fig. 5. The graph shows the correlation result of the participants' speech activity time series. The vertical red lines mark phase transitions which were annotated manually. The numbers mark the negotiation (I), presentation (II) and free phase (III) of the experiment where participants collaborated. The correlation changes during phase transitions where the focus shifts from the interlocutor to the experimenter or vice versa.

3.1 Structure in Verbal Dyadic Interaction

To better understand the distribution of gaps and the effect of erosion on the stability of utterance lengths we introduced a multiscale analysis of acoustic segment statistics. Specifically we coupled the histogram of segment length as a function of the erosion length d_p. Figure 4 depicts the result using a log color mapping for frequency, and showing d_p on the y-axis for a participant. Interestingly there are vertical bars at certain segment lengths, corresponding to repeated occurrences of specific utterance durations which remain quite stable under variation of d_p and gets more visible when pooled data of 20 participants is used. This (visually) suggests a corridor of $d_p \in [250, 1500]$ ms in which stable statements are rarely affected by the erosion approach.

3.2 Correlation

The processed data of both participants are used to calculate a windowed correlation as function of sample time k using equation (2).

$$\text{corr}_{xy}(k, \omega) = \frac{4}{\omega} \sum_{i=k-\omega}^{k} (sp_x[i] - 0.5) \cdot (sp_y[i] - 0.5) \tag{2}$$

We use a rectangular window function centered at $t = 0$. Local structure decreases with increased window size ω and stabilizes so that fast oscillations are filtered since the windowing operates similar to a low pass filter on the product feature $x \cdot y$. The correlation is computed on the *speech activity* feature introduced in section 3. Different from standard correlation, we shift the features so that silence is represented by $-\frac{1}{2}$ and speaking by $+\frac{1}{2}$. The motivation is that both joint silence and joint speaking should contribute in equal measure to positive correlation. For the correlation function to range between -1 and $+1$ the result is multiplied by 4.

4 Evaluation

To evaluate the correlation results, we manually annotated phases in our video data which occurred in a certain order in our experiment. We started with an *introduction*

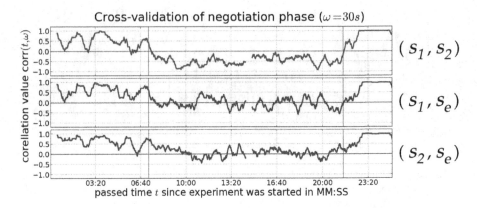

Fig. 6. The graphic shows the correlation for the time series s_1, s_2 from one trial and s_e from another trial. The vertical red lines mark the start and end of the negotiation phase which was annotated manually. The time series s_1 and s_2 (green graph) constantly anticorrelate after the phase transition until the end of the phase. Pre- and post-negotiation results of (s_1, s_e) and (s_2, s_e) look similar to the top graph, but steady anticorrelation cannot be observed during this phase. The gap at 13:40 is caused by different negotiation phase durations of the trials.

phase where the setting and task was introduced by the experimenter and the participants mostly listened or talked to a person from the experiment team. In the *negotiation phase* the participants had to discuss and agree on solutions for the recreation planning task. The study personnel left the room during that phase. The negotiation phase ended when the participants rang a bell and was followed by a presentation of the final solution. Between negotiation and presentation was a small window where the experimenter asked some question and handed out a questionnaire. After the presentation, the staff left the room a second time for about 5 minutes which was called the *free phase* where the participants were left sitting on the table without the mediating objects to record pure conversation data[4].

In Figure 5 both results are shown together for one trial exemplary. The correlation graph's zero crossing happens shortly after the negotiation and the presentation phase started and stays below zero right until the end of the phase. In the free phase we observe more fluctuation which additionally differs for every trial.

We cross-validated our findings by correlating time series from different trials to verify the approach. This was only done for phase transitions since these are essential moments for conversation detection and the phases' durations varied across the trials. Figure 6 shows such a cross-validation for start and ending of the negotiation phase. Speech activity time series s_1 and s_2 belong to the same trial that was shown in Figure 5 and were checked against s_e from another one. The blue graphs depict the inter-trial correlation and show similar shapes as the green graph before and after the end of the negotiation phase. During the phase the graph passes zero several times and fluctuates within the range of about $-\frac{1}{2}$ and $+\frac{1}{2}$ in both cases (s_1, s_e) and (s_2, s_e). The gaps are a result of the differing length of the negotiation phases.

---

[4] The participants were told that some system calibration had to be done to finish the experiment.

5 Discussion

Collaboration requires listening and a proper turn taking behavior where overlaps are accepted (in contrast to interrupts) and cause minor speech activity correlation. One person should speak at a time even though research has shown that there can be overlap towards the end of a turn or for backchannling (e.g. "yeah.. ahh") depending on the social norm, context and the interlocutors' relationship [5]. Weilhammer and Rabold found that average overlap related to the spoken language ranging from 150 to 330 ms for English, German and Japanese speakers [9].

The fact that cooperative speaking behavior anticorrelates is not surprising. But it is interesting how accurate this feature alone can determine if both participants cooperate. In our trials the probability of cooperative interaction was tightly coupled to the degree of the participants' verbal anticorrelation and its duration. Values smaller than $-\frac{1}{2}$ were hardly reached by cross-correlated time series.

For more fractured conversations this approach has to be adapted since the turn-taking time (also called inter-speaker interval) depends on the task [1]. The similarity of the inter-trial correlation with the intra-trial correlation shown in Figure 6 indicates that during those periods all participants, disregarding the trial, were listening most of the time to the experimenter's instructions which is supported by our qualitative analysis of the data. Joint silence is treated as uncooperative which is okay if both participants listen to the experimenter but it does not have to be true in all situations. We believe that this is one reason for fluctuation during the negotiation phase. Suppressing this behavior has to be done very carefully since in some cases the lack of verbal communication can be an indicator for recent problems in the problem solving process.

6 Conclusion

We have introduced and tested a new reliable signal-driven method for interaction focus detection from speech signal correlation. However, joint silence is treated as correlation and thus influences the current rating heavily. We propose a memory-based weight-decay feature to take the likeliness of a conversation between two (or more) interlocutors into account.

This approach may be useful to improve context awareness of future devices, a factor of increasing relevance in application development [6]. Importantly, this feature can be computed without any privacy-intrusion as no semantic features are accessed. Until full speech recognition-based interaction analsyis becomes available and cheap, our approach can support real-time situation detection.

As an interesting application beyond the scope of this paper we suggest the ubiquitous *Chatter Tracker* for parties, conferences or other social events: Every mobile phone running the application would collect speech activity to a server, which in turn computes pairwise correlations and composes for each interlocutor a summary of whom he has spoken with. Never forget to exchange contact information again as this could replace business cards.

Acknowledgement. This work has partially been supported by the Collaborative Research Center (SFB) 673 Alignment in Communication and the Center of Excellence for Cognitive Interaction Technology (CITEC). Both are funded by the German Research Foundation (DFG).

References

1. Bull, M., Aylett, M.: An analysis of the timing of turn-taking in a corpus of goal-oriented dialogue. In: Proceedings of ICSLP (1998)
2. Clark, H.: Using language, vol. 4. Cambridge University Press, Cambridge (1996)
3. Dierker, A., Mertes, C., Hermann, T., Hanheide, M., Sagerer, G.: Mediated attention with multimodal augmented reality. In: Proceedings of ICMI-MLMI, p. 245. ACM Press, New York (2009)
4. Dierker, A., Pitsch, K., Hermann, T.: An augmented-reality-based scenario for the collaborative construction of an interactive museum. Tech. rep., Bielefeld University (2011)
5. Edelsky, C.: Who's got the floor. Language in Society 10(3), 383–421 (1981)
6. Grudin, J.: The Computer Reaches Out: The Historical Continuity of Interface Design. In: Proceedings of CHI 1990, pp. 261–268 (1990)
7. Lecouteux, B., Vacher, M., Portet, F.: Distant Speech Recognition in a Smart Home: Comparison of Several Multisource ASRs in Realistic Conditions. In: Proceedings of Interspeech, pp. 2273–2276 (2011)
8. Wagner, D., Schmalstieg, D.: Artoolkitplus for pose tracking on mobile devices. In: Proceedings of CVWW (2007)
9. Weilhammer, K., Rabold, S.: Durational aspects in turn taking. In: International Congresses of Phonetic Sciences (2003)
10. Zehe, S.: BRIX - An Easy-to-Use Modular Sensor and Actuator Prototyping Toolkit. In: Proceedings of SeNAmI 2012, Lugano, Switzerland, pp. 823–828 (2012)

Server to Mobile Device Communication: A Case Study

Ricardo Anacleto[1], Lino Figueiredo[1], Ana Almeida[1], and Paulo Novais[2]

[1] GECAD, Knowledge Engineering and Decision Support Research Center,
School of Engineering of the Polytechnic Institute of Porto, Porto, Portugal
{rmsao,lbf,amn}@isep.ipp.pt
[2] CCTC - Computer Science and Technology Center,
University of Minho, Braga, Portugal
pjon@di.uminho.pt

Abstract. Develop a client-server application for a mobile environment can bring many challenges because of the mobile devices limitations. So, in this paper is discussed what can be the more reliable way to exchange information between a server and an Android mobile application, since it is important for users to have an application that really works in a responsive way and preferably without any errors. In this discussion two data transfer protocols (Socket and HTTP) and three serialization data formats (XML, JSON and Protocol Buffers) were tested using some metrics to evaluate which is the most practical and fast to use.

Keywords: Client-Server Communication, Mobile Applications, Protocol Buffers, Performance.

1 Introduction

Nowadays mobile devices still have several limitations (network traffic and battery consumption) compared to traditional computers that must be considered when developing a mobile application. It was based on these limitations that led us to the question: Which is the best way to exchange information between a server and a mobile client in order to minimize these limitations?

This question started to appear when developing a mobile application PSiS (Personalized Sightseeing Planning System) Mobile [1] to support a tourist when he is on vacations - more information about PSiS Mobile can be seen in section 2.

To answer this question a case study was performed, where the data transfer protocols performance were tested. It was done by transferring the points of interest data between the two sides (PSiS server and mobile application). Each point of interest is represented by 13 data fields where each one is formatted as string. The field which contains more data is the description, which in some cases can have more than 1000 characters. Each point of interest has about 600 Bytes of data.

Since the mobile application was developed to be used by an Android mobile device, a Google Nexus S with Android 4.1 was used. A normal notebook PC was used as server. Both were connected to the same IEEE 802.11g network. To decide which is the best technique to perform the data exchange, five metrics were used:

A. van Berlo et al. (Eds.): *Ambient Intelligence – Software & Applications*, AISC 219, pp. 79–86.
DOI: 10.1007/978-3-319-00566-9_11   © Springer International Publishing Switzerland 2013

- Process Duration, includes server request, data transfer, deserialization and data record on local database. This is important to realize which is the fastest technique;
- Average CPU load, important to see which system resources are being used;
- Average used Memory, the same as the previous one;
- Total bytes sent, this is very important because of the expensive data costs that carriers charge, less data consumption means less money spent;
- Total bytes received, has the same importance as the previous one.

In section 3 the performed case study is presented. This case study involves the transfer of points of interest from the server's database into the mobile device database using different technologies and based in some metrics to evaluate the results and understand which is the more appropriate to use. Section 4 presents an analysis and discussion about the obtained results. Finally, in section 5 some conclusions about the case study results are presented.

2   Case Study Context

The necessity to discover which is the best transfer protocol and data serialization format to transfer information between a server and a mobile application came when the authors were developing PSiS Mobile. This mobile application appears on the context of PSiS, which is a web application that aims to define and adapt a visit plan combining, in a tour, the most adequate tourism products (interesting places to visit, attractions, restaurants and accommodations) according to the tourists specific profile (which includes interests, personal values, wishes, constraints and disabilities) and available transportation system between different locations [1].

PSiS Mobile is composed by three pieces (see figure 1), the server-side, the middleware and the mobile client. In the server exists have a complete database with all the information about points of interest in a certain city/region and a complete users portfolio. The middleware was implemented to enable the communication between the server side and the mobile application.

The mobile client is a very important part of this system, because it is the bridge between the central services and the user visits. With a mobile device, the user can see the generated planning and the information about the nearby sights to visit, which are recommended according to his profile and current context. Also, the trip planning can be re-arranged according to the current context.

Since PSiS Mobile is an occasionally connected application, a temporary database is used on the mobile device to enable the access to part of the data without being constantly consuming network traffic, allowing the application to work without an internet connection (with some limitations, like no access to new points of interest).

After requesting a recommendation for a trip, all the necessary data is transferred from the server and stored on the mobile device. This was found to be necessary, because of the mobile Internet low speed rates and the possible unavailability. This necessary data represents the information about all the points of interest present on the planning schedule and other points of interest nearby the first ones.

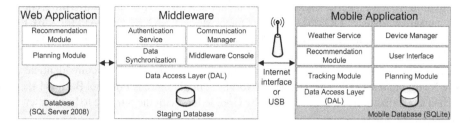

Fig. 1. PSiS Architecture Overview

3   Inter-process Communication Flow

There are several ways to exchange information between a server and a client, but in this case only two of the most used protocols were chosen to test, the Java Socket API [5] and the HTTP (Hypertext Transfer Protocol) REST (REpresentational State Transfer Web Services) [3]. The SOAP Web Services were left behind because of the bigger headers compared to the REST architecture, which increases the amount of network traffic and process power [7].

After data transfer protocols selection, the structure to serialize the information was defined. This is important in order to the two parties (server and client) "understand" each other, in this case it was chosen the XML, JSON and Protocol Buffers data structure formats.

Raw socket was the first tested approach since normally they are used to quickly exchange information [6]. First of all, a raw socket client and server modules were implemented. For each established connection, the server creates two threads: one to send data and another to receive data. Since there are two different threads the exchange can be performed asynchronously, avoiding waiting states on the client application. To test this protocol the data was serialized by the SAX Parser using a XML structure. With this protocol, message sizes were more compact since there aren't any headers (*e.g.*, HTTP or SOAP headers).

However, this system poses several problems in sockets management. Besides the need to specify a hard-coded and very inflexible communication protocol, raw sockets also need further implementation for error detection and transaction control.

The other tested protocol was HTTP, which is one of today's most popular client-server communication protocols. HTTP is a mature approach and a widely used protocol that already handle errors, simplifying its use and implementation. The only downside, comparing to the raw socket communication protocol, is the size of the sent/received data frames. This mainly happens because of the HTTP header, which is added to the sent/received data.

The header size along with the sent and received ACK (Acknowledgement) packages, to validate the transaction, varies between 6% and 10% of the size of the transferred data. For example, for a XML file with a size of 1.875 Mb, the client receives a total of 2.048 Mb (9% more than the original file size).

After the protocols chosen, three data structure formats were selected to test. The first one is the XML, since it is one of the most popular data structure formats used to store information. To have a better understanding about the XML performance three different XML parsers were used: DOM (Document Object Model), SAX (Simple API for XML) and Pull. DOM was chosen since it is the World Wide Web Consortium (W3C) standard and the other two because they claim to be the fastest XML files parsers.

Second one is JSON (JavaScript Object Notation) [2], which has a structure identical to the XML, but tries to be a low-overhead format. Finally there is Protocol Buffers [4], which is a serialization format developed by Google Inc. with the purpose to be simpler and faster than XML.

4   Empirical Analysis

In this section the results for each of the previously described exchange data techniques will be presented. To ensure more accurate results, four different tests with different file sizes were performed. Each of these tests was executed five times, and the presented results are the average of the five attempts. The file sizes, for each test and data serialization format are described on table 1.

Table 1. File sizes (in kB) for each test and data serialization format

Serialization format	First	Second	Third	Fourth
XML	1	253	375	1875
JSON	0.779	227	313	1564
ProtocolBuffers	0.665	195	256	1276

In the first test only the information of one point of interest was used. This was valuable to get a first look of the mobile devices behavior when few data bytes are exchanged over network compared to big files.

Analyzing table 2, it appears that the fastest architecture is HTTP using Protocol Buffers, followed by HTTP using JSON. The raw socket protocol was slower mainly because of the connection initialization, which is a time consuming process, especially when we try to detect and control communication errors.

However, as expected, it was the raw socket with XML architecture that had fewer bytes transferred between server and client followed by the HTTP protocol with Protocol Buffers. Finally, HTTP with XML is the heaviest of them all.

To this test weren't provided any data for the CPU load and memory metrics because the process is completed so quickly that significant values can't be obtained (the readings are made per second).

In the second test the information about 250 points of interest was transferred. One of the most relevant findings is that the XML parsing algorithms have significant performance differences. The DOM, one more time, was the slowest and SAX proved to be the fastest, surpassing Protocol Buffers that only in this test wasn't the best.

Table 2. First test results

Protocol	Duration (ms)	CPU (%)	Memory (MB)	Data Received (kB)	Data Sent (kB)
HTTP XML SAX	595	-	-	1.5	0.5
HTTP XML DOM	773	-	-	1.5	0.5
HTTP XML Pulll	555	-	-	1.5	0.5
HTTP JSON	511	-	-	1.2	0.5
HTTP PROBUF	**506**	-	-	1.1	0.5
SOCKET	1893	-	-	**1.0**	**0.5**

Table 3. Second test results

Protocol	Duration (ms)	CPU (%)	Memory (MB)	Data Received (kB)	Data Sent (kB)
HTTP XML SAX	**2023**	46.5	4.59	270.0	5.8
HTTP XML DOM	14947	90.1	6.02	270.2	5.5
HTTP XML Pulll	4940	76.4	5.43	270.0	7.1
HTTP JSON	3784	78.9	5.26	241.2	6.4
HTTP PROBUF	2036	55.8	5.03	**206.9**	5.3
SOCKET	7485	**22.7**	**4.27**	262.9	**4.7**

Looking at table 3, can be seen that socket method consumes less system resources (CPU and memory) than the others because it doesn't have so many parsing routines. However, the whole process still takes a long time to execute. Protocol Buffers was the one that had transferred less bytes, since it includes some data compression.

In the third test, it was transferred the information about 461 points of interest. The results follow the same pattern of the previous tests, where Protocol Buffers was the fastest, though only for a little margin (table 4).

JSON behaved as expected, serialization turns the file lighter than XML, but it has a weak decoder (the Android platform native JSON parser was used) and becomes slower when compared with the, also Android native, SAX Parser.

Analyzing the CPU utilization data, can be observed that the worst is HTTP with DOM parser, since it uses an average of 93% during 26 seconds, which can represent a lot of battery spent. Another important analysis is that the socket method only has used 48% of CPU but it has an overall duration of almost 8 seconds. Comparing it with HTTP using Protocol Buffers, can be seen that Sockets aren't so good, because HTTP with Protocol Buffers uses 51% but only for 2 seconds. Considering the memory usage, socket method uses less memory than the others protocols.

Finally, the fourth test, where it was decided to perform a more thorough test to denote additional differences on the obtained results. In this test the information about 1884 points of interest (four times all the points of interest stored on the database) was used.

Table 4. Third test results

Protocol	Duration (ms)	CPU (%)	Memory (MB)	Data Received (kB)	Data Sent (kB)
HTTP XML SAX	2797	70.4	5.65	398.9	8.6
HTTP XML DOM	26985	93.3	5.95	399.4	8.5
HTTP XML Pulll	5331	81.6	5.44	398.9	8.4
HTTP JSON	4876	79.7	5.16	332.6	8.6
HTTP PROBUF	**2316**	51.0	5.14	**271.5**	7.4
SOCKET	7949	**48.0**	**4.99**	384.7	**7.0**

Table 5. Fourth test results

Protocol	Duration (ms)	CPU (%)	Memory (MB)	Data Received (kB)	Data Sent (kB)
HTTP XML SAX	12171	**70.0**	6.59	2048	37.3
HTTP PROBUF	**10060**	72.3	**5.89**	**1400**	**28.5**

Comparing the third with the fourth test, can be observed that the processing time has been 5 times more and the amount of data transferred is only 4 times the transferred data on the third test. This is mainly explained because of the limited mobile device memory. The operating system is always trying to get more and more memory and it slows down the entire process.

Notice that only results for two techniques are provided. This happened because all the others gave an "Out of Memory" error due to the mobile device lack of memory. This happens because Android heap memory is limited to 16MB per application on the most available devices, and only the high-end ones have a limit of 24MB. These two techniques were also the ones that have produced better results in the other tests (HTTP with SAX Parser and HTTP with Protocol Buffers). As can be seen on table 5 both used almost the same system resources.

In this test can be seen a bigger difference in performance between Protocol Buffers and SAX, especially in the transferred data size. Protocol Buffers transmitted about 600 kB less data (since the serialized file is that much smaller) and in lesser two seconds than the SAX parser.

5   Conclusions

The purpose of this study was to discover which technology/technique is more reliable and faster to use in a server-Android mobile application environment. Therefore, in this chapter the conclusions about the obtained results and what technique was chosen to use are presented. Also, some considerations that have been learned and validated during these tests are discussed.

In theory, socket approach seems to be the right choice. In practice, it was found some important disadvantages compared to the other approaches, since it proved to be error prone and slower. Considering the analysis of cost over benefit between this approach and HTTP, it was concluded that the socket gains on the transferred kBs between the two sides, don't outweigh the associated disadvantages. The socket results can be explained by a poor optimization of the Android Socket API.

Sockets were left behind due to the few advantages that they actually bring, compared to the HTTP protocol. Also, raw sockets are much more complex and hard to work with. It's like reinventing the wheel when it already exists. On the other hand, HTTP is reliable and is able to perform error handling. HTTP was the chosen protocol for the PSiS Mobile implementation. With these tests the research team attested, that the time spent in the implementation of sockets is not worth the supposed superiority of performance, which in this case there wasn't any of it besides the smaller data messages.

After choose the transfer protocol the most commonly used data serialization formats to encapsulate the data to be sent over that protocol were inspected. Starting with XML, the case study revealed that after all it isn't so slow to parse, but instead it depends highly on the used parser. Regarding file size it is only slightly behind the others, because of the inclusion of multiple tags and for no data compression implementation. Another issue that has to be considered is to not rely only in the theory, but try to understand it and put it into practice in order to confirm the results for our case.

Considering the XML parsers, it is noteworthy that DOM is definitely the slowest and the most complex to work. The SAX ends up having a similar performance to Protocol Buffers, which proved to be the lightest and the fastest in almost all the tests. These two are, according to our tests, the best approaches. SAX is overtaken by the Protocol Buffers when it comes to speed and file sizes, thus can be concluded that Protocol Buffers is the fastest and lightest serialization format. Then and as expected, since it is one of its claims, JSON files are smaller. However, the Android native JSON parser proved to be slower than the best XML parser.

According to the previous statements, the HTTP protocol in conjunction with Protocol Buffers was the chosen mechanism to exchange information between PSiS server and mobile application, since it spent less system resources (therefore less battery) and less network data consumption. Thus, some of the limitations of mobile devices were minimized.

Another lesson that was learned is that there is no advantage in sending few or a lot of information at once, but something in between them. If few information is sent at once a great waste of time exists in the initialization of the communication. Comparing the second and third tests, where twice the information was sent, can be seen that it takes just a little more time to process it. However, if a lot of information is sent at once, as done in the fourth test, some memory problems can be experienced and thereby slow down the whole process. The best thing to do is to choose something in the middle, *i.e.*, medium-sized files.

Finally, the research team has learned that it is worth investing some time in these small tests, because with them the user experience can be improved. These tests don't take so long to implement and can result in a good knowledge for the team. Has can be seen, for Android platform the HTTP protocol and Protocol Buffers are well implemented and it is worth to give a try, getting a fast and reliable solution to transfer information between a server and an Android mobile device.

Acknowledgement. The authors would like to acknowledge FCT, FEDER, POCTI, POSI, POCI and POSC for their support to GECAD unit, to the project PSIS (PTDC/TRA/72152/2006) and for the PhD grant (SFRH/BD/70248/2010).

References

1. Anacleto, R., Luz, N., Figueiredo, L.: Personalized sightseeing tours support using mobile devices. In: Forbrig, P., Paternó, F., Mark Pejtersen, A. (eds.) HCIS 2010. IFIP AICT, vol. 332, pp. 301–304. Springer, Heidelberg (2010)
2. Crockford, D.: JSON: the fat-free alternative to XML. In: Proc. of XML, vol. 2006 (2006)
3. Fielding, R.T., Taylor, R.N.: Principled design of the modern web architecture. ACM Transactions on Internet Technology (TOIT) 2(2), 115–150 (2002)
4. Google: Protocol buffer (2012),
   http://code.google.com/apis/protocolbuffers/docs/overview.html
5. Harold, R., Loukides, M.: Java network programming. O'Reilly & Associates, Inc., Sebastopol (2000)
6. Pakin, S., Karamcheti, V., Chien, A.A.: Fast messages: Efficient, portable communication for workstation clusters and MPPs. IEEE Concurrency 5(2), 60–72 (1997)
7. Pautasso, C., Zimmermann, O., Leymann, F.: Restful web services vs. big'web services: making the right architectural decision. In: Proceeding of the 17th International Conference on World Wide Web, pp. 805–814 (2008)

Zappa: An Open Mobile Platform
to Build Cloud-Based m-Health Systems

Ángel Ruiz-Zafra, Kawtar Benghazi, Manuel Noguera, and José Luis Garrido

Dpt. Lenguajes y Sistemas Informáticos,
University of Granada, E.T.S.I.I., c/Saucedo Aranda s/n, 18071 Granada, Spain
{bihut,benghazi,mnoguera,jgarrido}@ugr.es

Abstract. Cloud computing and associated services are changing the way in which we manage information and access data. E-health services are not impermeable to novel technologies, especially those that involve mobile devices. At present, many patient monitoring m-health (mobile-health) platforms consist of close, vendor-dependent solutions based on particular architectures and technologies offering a limited set of interfaces to interoperate with. This fact hinders to advance in quality attributes such as customization, adaptation, extension, interoperability and even transparency of cloud infrastructure of existing solutions according to the specific needs of their users (patients and physicians). This paper presents an extensible, scalable, highly-interoperable and customizable platform called Zappa, designed to support e-Health/m-Health systems and that is able to operate in the cloud. The platform is based on components and services architecture, as well as on open and close source hardware and open-source software that reduces its acquisition and operation costs. The platform has been used to develop several remote mobile monitoring m-health systems.

Keywords: m-Health, e-Health, mobile applications, patient monitoring, SOA, open source, cloud computing.

1 Introduction

The appearance of mobile systems for e-Health has revolutionized this domain by providing tools that enable, among others, remote monitoring of patients and automate both the capture of health parameters data and its transference to remote computer systems, thereby, giving rise to so called m-health systems [1]. Furthermore, m-health systems allow patients to be under steady medical supervision so that changes in patients' health status can be tracked and notified. Unattended care is also possible, i.e., peak values in certain parameters and temporary crisis that alert of abnormal situations can be registered, which otherwise might go unnoticed, thereby affecting patients' health due to the lack of fairly relevant information in his/her medical record [2].

The challenge is not only to automate these tasks, but to also try to assure certain quality attributes of the system that support them through the application of design techniques. For instance, although there exist several monitoring systems that provide plenty of functionalities, in many ways these systems do not cover the different needs

A. van Berlo et al. (Eds.): *Ambient Intelligence – Software & Applications,* AISC 219, pp. 87–94.
DOI: 10.1007/978-3-319-00566-9_12   © Springer International Publishing Switzerland 2013

of a wide range of patients, since each patient requires a personalized and adaptive monitoring to be defined by physicians [3].

Likewise, physicians demand technological infrastructures that provide seamless interaction with medical systems any-where anytime through internet-operable clouds where the clouds themselves are a transparent element.

This paper presents the design of an extensible, scalable and customizable cloud platform for the development of eHealth/mHealth systems. Some key design decisions are based on the use of the recent concept for delivering resources as services over Internet (Cloud computing), open technologies (open-source software, open hardware, etc.) and additional techniques, for instance, the personal information about patients who use the platform is to be remotely managed in a customizable manner. Moreover, the platform is intended to provide uninterrupted monitoring with the goal of obtaining some information that can be subsequently analyzed by physicians for diagnosing.

This paper is organized as follows. Section II presents related work. Section III presents the platform and its features. Section IV introduces software applications and tools based on the platform. Finally, Section V summarizes the conclusions and future work.

2   Related Work

So far, several mobile systems for patient monitoring have been proposed. For space limitation, it is not possible to compare the present proposal with each one of them. Nonetheless, one common drawback of many of them is that they are designed with hardly-scalable architectures and implemented with privative software [6][10]. This not only makes it difficult for other developers to work with them, but also increases their cost and limits their adaptability to be customized for a wide range of patients [7]. In other cases, the set of services provided is limited or focused on specific services, such as information representation/visualization [4], just show the health status of one vital sign in an exact moment of time [5], are designed and devised for local monitoring of patients and, limit the number of patients that can make use of the system [12][13], or monitor just one vital sign [14][15].

3   Zappa: Cloud m-Health Platform

In the following sections it is introduced the Zappa platform, the intended properties and principles that have guided its design, as well as the constituents of its system architecture.

3.1   Desired Features

E-health and m-health platform must comply with certain desired features that foster their acceptance by the different group of users that interact with them (patients, relatives, physicians, etc.). The proposed platform aims at providing certain monitoring services for medical purpose and intends to achieve the following characteristics:

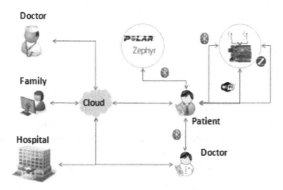

Fig. 1. Local-Remote monitoring scenario

1. System adaptability, which allows the customization of the application in order to meet heterogeneous and changing patients' needs. Integration of user profiles that are used to set up the application (i.e., determine the biosensors to be used, peak of values for triggering alerts, etc.). This information is internally represented in XML to foster interoperability.
2. Timely alert, when a peak value in certain patient clinical parameter is registered, the system should react in the predefined interval of times to safely notify about abnormal situation to professionals and caregivers. Besides, this value will be stored in the patient medical record.
3. Self-configuration, the system which must adapt to changing needs of clinical conditions of patients, their profiles, and their dynamic environment, without external tools or expert knowledge.
4. Cost-reduced system, by making use of:
   - Open Source software, so that system modules can be reused by developers and extended without paying for licenses, and thus, reduce costs.
   - Open Hardware, which required the different elements to be based on open hardware, to reduce the costs for customers [3][8].
5. Dynamic System. M-health scenarios are dynamic in nature. This means that not only new components (software components and hardware devices, like biosensors) are able to automatically connect and disconnect to the system at run-time, but also new patients can become part of a scenario and interact with the system on the fly. Fig. 1 shows a generic local-remote scenario where the patient is being monitored locally and remotely (family, doctor, hospital).
6. Interoperable system, in order to overcome the heterogeneity of system devices and enable the system to communicate and integrate with other systems. XML-based representations of data in conjunction with some extensible and adaptable communication protocols are used to achieve this goal.
7. Seamless clouds, i.e., users (either physicians or patients), are not aware that they are interacting with a cloud infrastructure.
8. Use of medical devices. The system should be able to use different medical devices based on different technologies. This feature guarantees the use of commercial and

close-source hardware devices of popular vendors, such as Polar or Zephyr (Bluetooth), and custom/research devices based in open-source hardware as Arduino (ZigBee, Bluetooth, Wifi).

3.2   Architecture

According to the characteristics enumerated in the previous section different concerns of mHealth systems can be identified, e.g., communication with and between devices, data representation, formatting and storage, wireless protocols, cloud access, etc. These concerns are supported by means of a component-based (modular) architecture design so that changes in one system component do not alter system behavior and do not affect the configuration of other system components either [11]. Fig. 2 shows the Zappa platform

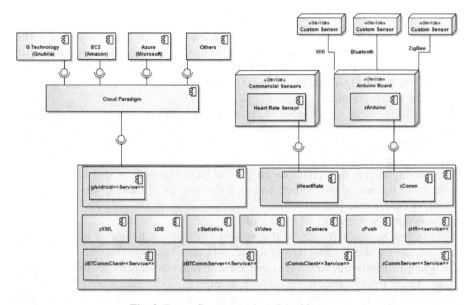

Fig. 2. Zappa Component-based Architecture

architecture. There exist different components that encapsulate and abstract all functionalities related to information management (e.g., zXML, zDB), communications (zComm, ZBTServer) and cloud technologies (gAndroid for G Technology), as well as biosensor devices, among others. These components could be used in eHealth/mHealth systems (Android). Some design decisions, such as the management of the communication between different devices with different components(zComm for Wifi and zHeartRate for Bluetooth) or separate the cloud management in different components, have been adopted in order to foster higher extensibility of the mechanisms and functionalities the platform permits to:

1. Connect a new biosensor or disconnect a biosensor at run-time. The status of biosensors will be detected and updated automatically, thanks to zComm (for custom biosensors) and zHR (for Bluetooth heart rate sensors) components.

2. Establish the communication between two devices (i.e., between two mobile devices or between a mobile device and a biosensor) with any of usual communication protocols, like Bluetooth and Wifi. This is achieved using different components (zHeartRate with Bluetooth and zComm/zArduino with Wifi) which are used in services that are executed in the background to provide a full time support. Thanks to message passing communication protocols implemented in zComm and zArduino components, the developer can add new functionalities and specify new types of notification messages, without changing the behavior or performance of the system.

3. Interact with cloud infraestructures. For example, the gAndroid component has been explicitly designed to work with G-Technology IaaS (Infrastructure as a Service). If interoperation with other IaaS vendors is required, new components can be added to the platform seamlessly.

4   Example Applications

In order to show the applicability of the platform, several m-health applications based on the Zappa platform are introduced. These applications use different components (Fig 2) of the platforms to provide the corresponding functionality.

4.1   Zappa App

Zappa App is an m-Health system used to monitor the heart rate, temperature and blood pressure of the patient. In addition, the system is able to save the vital sign values, detect health problems and share information with a doctor or medical staff that are in the same place that the patient (Bluetooth).

Currently, Zappa App consists of four software applications. Three of these applications have been built for mobile devices (Android) and the last one (desktop application) has been built to support the information exchange between mobile devices and the biosensors connected to Arduino boards (an open-source electronics prototyping platform based on flexible, easy-to-use hardware and software), which makes use of the Zigbee technology for communication. The different mobile applications are the following:

— Setting Application (Zappa App Setting): It is the application installed on the patient's mobile devices. It is used to manage all the different elements of the system. It allows the user to manage sensors (zComm component), statistics (zStatistics component), profiles and patients (Fig. 3 - left).

— Monitoring Application (Zappa App Patient): This application (Fig. 3 - center) shows the obtained values from the biosensors through the communication service (zComm and ZCommService components). Each monitorization is customized and varies depending on the selected user profile. This application has been built as an Android service. In this manner the system (alert notifications, real-time biosensors connection/disconnection, information storage and other tasks, can be executed even when the application is running in background. In addition, this application starts the different components that allow sending and receiving information from the cloud. In this manner, the application can receive requests and act accordingly.

— External request application (Zappa App M.D.). This application could be installed in any mobile device and it able to search, via Bluetooth, different patients in the same location and to obtain personal in-formation or monitoring values. This is possible using zBTCommClient component in this app, and the component zBTCommServer in the mobile device of the patient (Fig. 3 -right).

Fig. 3. Zappa App

4.2   Cloud Rehab

Cloud Rehab is a full m-Health system that is used to monitor the daily activities of patients with severe brain damage. This kind of patients can evolve faster performing certain activities, such as having lunch, dressing, etc., if these activities are monitored (in real-time) by physicians or medical specialist. The doctors can define new activities, called sessions, according to the evolution of the patient.

One session is made up of a training video of a patient performing an activity in a training process, a set of audio and image files used when the heart rate value of the patient reach a certain value, the recorded video of the session, the heart rate values, and alert triggered, among others. The doctor can define new sessions with new training activities and, once the patient has completed them, review the session's information to evaluate patient progress. The system includes two applications.

– Web Application: The web application (Fig. 4 left) is used by the medical staff to manage patients' medical information. This application is used to manage sessions, monitor patients values, real-time monitoring, etc. The application can be also used by the relatives of the patient to check if she/he has completed all the sessions, her/his medical progress and/or check if the doctor has defined a new session.
– Android Application: This application (Fig. 4 right) is used by the patient to perform the sessions. The application monitors heart rate in background (zHR component as a service) and stores the values in a cloud server (gAndroid component as a service) and in a local database in real-time (zDB component). If the heart rate value is above or below some predefined thresholds, the patient can use relaxation options.

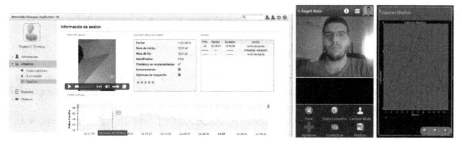

Fig. 4. Cloud Rehab Web Application (left) and Android Application (right)

5   Conclusions and Future Work

Health monitoring systems can help patients in many different ways [3]. Normally the control of the status health of the patient is good enough, but thanks to mobile devices and their hardware and software evolution it is possible to provide more functionality to them like remote at home monitoring.

An m-health cloud-transparent platform, called Zappa, for developing systems intended to monitor patients remotely and in real time has been described. Its design is based on an open architecture and made up of components based on standard technologies that allow a higher extensibility and support the management of different patients, being able to customize the systems to be developed to their needs. The platform aims also at saving implantation and economic cost through open source hardware and software. The platform enables systems to operate in hybrid monitoring scenarios and offers robust communication capabilities by enabling to switch between different protocols seamlessly and on the fly depending on availability, efficiency, etc., issues.

Zappa supports the construction of m-health cloud-based systems so as to provide functionalities fulfilling non-functional properties that foster patient acceptance of such systems. Several sample applications have been shown in order to illustrate the applicability of the proposal.

We are currently working in the improvement of the proposed platform. Plans for future work include: 1) the development of an expert system, so as to improve diagnosing; 2) the improvement of the main web server, in order to incorporate new functionalities such as external remote calls to public emergency services (112, 911, etc.), security and privacy control; and 3) the development of additional components to adapt the platform with other cloud platforms such as EC2 or Windows Azure will be designed and developed in the future.

Acknowledgements. This research work has been funded by the CEI BioTIC Granada under project 20F2/36, Innovation Office from the Andalusian Government under project TIN-6600 and the Spanish Ministry of Economy and Competitiveness under the project TIN2012-38600.

References

1. Mechael, P.N.: The Case for mHealth in Developing Countries. MIT Press Journals - Winter 4(1), 103–118 (2009)
2. Siebra, C., Lino, N., Silva, M., Siebra, H.: An Embedded Mobile Deductive System for Low Cost Health Monitoring Support. In: 24th International Symposium on Computer-Based Medical Systems (CBMS), pp. 1–6 (2011)
3. Matheou, G., Kyriacou, E., Chimonidou, P., Pattichis, C., Lambrinou, E., Barberis, V.I., Georghiou, G.P.: A Post Cardiac Surgery Home-Monitoring System. In: CompSysTech 2011, Proceedings of the 12th International Conference on Computer Systems and Technologies, pp. 341–346 (2011)
4. Jovic, A., Bogunovic, N.: HRVFrame: Java-Based Framework for Feature Extraction from Cardiac Rhythm. In: Peleg, M., Lavrač, N., Combi, C. (eds.) AIME 2011. LNCS, vol. 6747, pp. 96–100. Springer, Heidelberg (2011)
5. López, G.V.: Construction of a signal acquisition equipment for use with sensors for biomedical applications
6. http://www.airstriptech.com/
7. Patel, S., Lorincz, K., Hughes, R., Huggins, N., Growdon, J., Standaert, D., Akay, M., Dy, J., Welsh, M., Bonato, P.: Monitoring Motor Fluctuations in Patients With Parkinson's Disease Using Wearable Sensors. IEEE Transactions on Information Technology in Biomedicine 13(6) (2009)
8. Akter, S., D'Ambra, J., Ray, P.: Service quality of mHealth platforms: development and validation of a hierarchical model using PLS. Electronic Markets 20(3-4), 209–227
9. Korhonen, I., Parkka, J., Van Gils, M.: Health monitoring in the home of the future. In: IEEE Engineering in Medicine and Biology Magazine, Dept. of Res., VTT Inf. Technol., Tampere, Finland, vol. 22(3), pp. 66–73 (2003)
10. Villalba, E., Salvi, D., Ottaviano, M., Peinado, I., Arredondo, M.T., Akay, A.: Wearable and Mobile System to Manage Remotely Heart Failure. IEEE Transactions on Information Technology in Biomedicine 13(6), 990–996 (2009)
11. Zenger, M.: KERIS: evolving software with extensible modules. Journal of Software Maintenance 17(5), 333–362 (2005)
12. Żwan, P., Kaszuba, K., Kostek, B.z.: Monitoring Parkinson's Disease Patients Employing Biometric Sensors and Rule-Based Data Processing. In: Szczuka, M., Kryszkiewicz, M., Ramanna, S., Jensen, R., Hu, Q. (eds.) RSCTC 2010. LNCS, vol. 6086, pp. 110–119. Springer, Heidelberg (2010)
13. Jeroslav, J., Jan, K.: Mobile personal system for monitoring ill and endangered people (Medical Personal Watcher). In: International Workshop on Wearable Micro and Nanosystems for Personalised Health (2008)
14. Uribe, C., Isaza, C., Florez-Arango, J.F.: Qualitative-fuzzy decision support system for monitoring patients with cardiovascular risk. In: 2011 Eighth International Conference on Fuzzy Systems and Knowledge Discovery (FSKD), pp. 1621–1625 (2011)
15. Gangwar, D.S., Saini, D.S.: Arrogyam: Arrhythmia detection for ambulatory patient monitoring. In: Ranka, S., Banerjee, A., Biswas, K.K., Dua, S., Mishra, P., Moona, R., Poon, S.-H., Wang, C.-L. (eds.) IC3 2010. CCIS, vol. 95, pp. 168–180. Springer, Heidelberg (2010)

Elicitation of Quality Characteristics
for AAL Systems and Services

Aida Omerovic, Anders Kofod-Petersen, Bjørnar Solhaug, and Ingrid Svagård

SINTEF ICT, Trondheim, Norway
{Aida.Omerovic,Anders.Kofod-Petersen,
Bjornar.Solhaug,Ingrid.Svagard}@sintef.no

Abstract. Ambient Assisted Living (AAL) is a promising and fast growing area of technologies and services to assist people with special needs (e.g. elderly or disabled) in managing more independently their everyday life. AAL is founded on increasing needs for welfare technologies, as well as on significant effort from many scientific disciplines, the society, and the industry. The research has so far been primarily concentrated on elicitation of the functional aspects and on providing the technical solutions for the AAL systems and services. The problem of eliciting non-functional requirements and quality characteristics that are specific and critical for AAL, however, has been addressed to a much lesser extent. Failing to ensure the necessary system and service quality regarding critical characteristics may represent a significant obstacle to the wider acceptance of AAL in the society. There is hence a need to increase awareness of quality of AAL systems and services by providing the necessary supplement to the established state of the art. This paper reports on the process and the results from elicitation of AAL specific quality characteristics. The approach is based on established reference architectures and roadmapping material, as well as the ISO/IEC 9126 software product quality standard. The paper demonstrates how to do the elicitation in practice, and proposes the set of quality characteristics that are most important in the AAL context.

Keywords: Ambient Assisted Living, AAL Systems and Services, Quality Characteristics, ISO/IEC 9126.

1 Introduction

Ambient Assisted Living (AAL) [8,14] is a growing application domain of Ambient Intelligence [3], and is characterized by responsive ICT systems that are used for empowering people with special needs for managing their everyday activities. The goal is to make the life of the end users more comfortable and independent, supporting them in maintaining an active and creative participation in the community and in their preferred living environment. Due to the societal and demographical changes in the society of today, the need for welfare technologies in general and AAL systems and services in particular, is growing. This is also clearly recognized at EU level with the Horizon 2020 funding programme [6], where health, demographic change and wellbeing constitute one of the identified focus areas, and ICT is stressed as one of the key enabling technologies. Moreover, significant effort from many scientific disciplines and the industry is driving the development of the AAL technologies and application areas.

A. van Berlo et al. (Eds.): *Ambient Intelligence – Software & Applications*, AISC 219, pp. 95–104.
DOI: 10.1007/978-3-319-00566-9_13   © Springer International Publishing Switzerland 2013

The AAL research and development activities have so far primarily concentrated on elicitation of the functional aspects, as well as on providing the technical solutions for the AAL systems and services [16]; there are few approaches on how to do a systematic elicitation of the AAL-specific non-functional characteristics. Such characteristics should be taken into account already during development as inherent aspects of the very architecture and design of AAL systems. It is crucial for the acceptance of the AAL technologies that they are able to fulfill and verify both functional and non-functional requirements, taking into account all relevant stakeholders.

The main contribution of this paper is twofold. First, we propose an approach to elicit, weight and document non-functional requirements in terms of so-called quality characteristics. The approach is based on international standards on software product quality, as well as AAL reference architectures and roadmapping as proposed by leading communities in the field. Second, we report on our results of using this approach by documenting the most important AAL quality characteristics. These characteristics can be understood as a necessary and adequate complement to existing reference architectures, and serve as guidance for service developers and providers.

The rest of the paper is organized as follows. In Section 2 we describe the underlying basis for our work, both on AAL and on product quality, and we describe our methodological approach. In Section 3 we describe in more details the process undergone for how to do the elicitation of the quality characteristics, and in Section 4 we present the results. In Section 5 we discuss the threats to validity and reliability, addressing in detail the particular uncertainties related to the process undergone and the results. Related work is briefly summarized in Section 6, before concluding and proposing directions for future work in Section 7.

2   Background

The specification and evaluation of software product quality requires the product to be at hand, either as a requirements statement, as a design document, or as an implementation. When addressing AAL systems and services in general, as in this paper, there are basically two ways of approaching quality specification and evaluation, namely bottom-up or top-down. The bottom-up approach is to select a set of representative AAL systems and define the quality characteristics for each of them, before consolidating the results. The top-down approach is to use a generic specification of AAL systems and define the quality characteristics that are appropriate at this level. These characteristics should then be adequate for all instances of the generic specification.

In our work we have used the top-down approach by carefully selecting AAL documentations that are developed by leading communities, that span most application domains, and that serve as a common abstraction of typical use cases and system instances. First, at the level of AAL architecture, we have adopted the reference architecture [13] of the universAAL FP7 research project [12]. Generally, the purpose of a reference architecture is to generalize and extract common functions and configurations, and to provide a base for instantiating target systems that use that common base more reliably and cost effectively [15]. Complementing a reference architecture with adequate quality characteristics should serve the same purpose for such non-functional requirements.

Second, at the level of use cases, usage areas, users and recent driving developments within AAL, we have used the AALIANCE Ambient Assisted Living Roadmap [14] which is a comprehensive roadmap and strategic guidance for R&D approaches in the AAL context.

As a basis for the elicitation of the adequate quality characteristics, we used the standardized ISO/IEC 9126 series on product quality in software engineering [9]. ISO/IEC 9126 provides an established specification of decomposed quality notions with their qualitative and quantitative definitions. The clear advantage of this international standard for expressing quality characteristics, is the possibility of comparing one system with another. The standard defines a quality model for *external and internal quality*, and for *quality in use*. External quality is the totality of the characteristics of the software product from an external view when the software is executed. Internal quality is the totality of characteristics from an internal view and is used to specify properties of interim products. The characteristics of the internal and external quality model are functionality, reliability, usability, efficiency, maintainability and portability. These are in turn decomposed into a total of 34 sub-characteristics. Quality in use is the user's view of the quality of the software product when it is used in a specific environment and a specific context of use. The quality in use characteristics are effectiveness, productivity, safety and satisfaction.

Determining the adequate quality characteristics for a specific product obviously depends on who the relevant stakeholders are. In our approach we have aimed for the identification of generic quality characteristics independent of the concrete stakeholders in the specific AAL system instances. However, the quality characteristics should capture the requirements and expectations of AAL service providers and consumers; quality in use should in particular ensure that users can achieve their goals in a particular environment [9].

To ensure as much as possible an objective and unbiased quality elicitation, the process was conducted in two independent strands by two separate groups. One group consisted of experts from the AAL domain who conducted the elicitation based on their experience and expertise, and by actively using the universAAL reference architecture [13] as the normative specification of AAL systems and services. We refer to this strand as Domain Expert Judgment (DEJ) based process.

The other group consisted of experts on security and quality assessment who conducted the elicitation by a thematic analysis [7] of the AALIANCE Roadmap [14]. Thematic analysis is one of commonly used methods of qualitative research analysis. In thematic analysis, data are explored and coded according to patterns or commonalities (themes). The codes emerge as the data are examined. Then, the codes are analyzed by comparing theme frequencies and the relationships between the themes. We refer to this strand as Thematic Analysis (TA) based process.

Both the DEJ-based elicitation and the TA-based elicitation resulted in a specification of all ISO/IEC 9126 quality characteristics with weights and their rationale in the AAL context. As explained in more details in the next sections, the respective results of the two groups were subsequently compared and consolidated.

3   Process

This section reports on the process of elicitation of the AAL-specific quality characteristics. In order to ensure, as much as possible, an unbiased and objective process, we combined two independent approaches, namely the DEJ-based process and the TA-based process. Based on two parallel processes, we have instantiated the ISO/IEC 9126 software product quality standard [9].

Four participants were involved in the elicitation process. Two of them (participants A and B) are experts in AAL and researchers in the fields of welfare technologies and ambient intelligence, respectively. The other two (participants C and D) are researchers in the fields of model-based security and quality analysis. Each participant had at least ten years of professional experience in the respective domain of expertise. Participants A and B performed the DEJ-based elicitation, and participants C and D performed the TA-based elicitation.

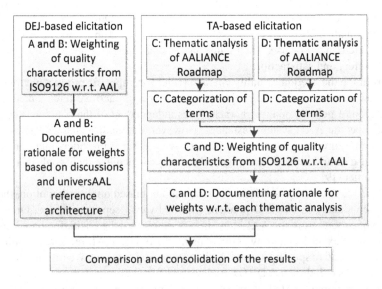

Fig. 1. The two independent approaches in the process of the quality characteristics elicitation through instantiating the ISO/IEC 9126 standard, followed by comparison and consolidation of the results

The instantiation, which was performed independently by the two groups, involved weighting of each quality characteristic with one of three possible marks (High, Medium, or Low) and providing a rationale for the evaluation, as proposed by the ISO/IEC 9126. All the four participants were familiar with the standard beforehand and actively used it during the marking process, in order to retrieve the correct definitions of the quality characteristics. As illustrated by Figure 1, the process undergone involved the following steps during the two above mentioned independent directions:

- **TA-Based Elicitation:** AAL quality characteristics elicitation by participants C and D. This part of the process involved a thematic analysis of the above mentioned AALIANCE Roadmap [14] with respect to quality aspects. The AAL quality related contents extracted were coded and the codes were categorized in the form of overall quality characteristics. The two participants independently obtained 8 and 7 categories, respectively. The underlying codes for each category were documented. The number of categories was not pre-determined and the possible categories were not pre-defined. Based on the codes (i.e. their frequency of occurrences) and categories obtained, the quality characteristics from ISO/IEC 9126 software product quality standard were weighted. While the thematic analyses by the two participants were performed independently, the weighting was made during a joint session. The rationale was documented with references to the results of the thematic analyses.
- **DEJ-Based Elicitation:** AAL quality characteristics elicitation by participants A and B. This part of the process involved expert judgments in weighting quality characteristics from ISO/IEC 9126 software product quality standard with respect to AAL in general and to the universAAL reference architecture [13] in particular. This was performed in the form of a workshop and the rationale for each weight was documented.

The comparison and consolidation step was an overall consideration of all results by all participants, where an agreed weight was assigned to each quality characteristic. The weights which had full matches needed no further consolidation, while the ones that deviated between the two elicitation approaches were consolidated by comparing the rationales and agreeing upon a single weight.

4   Results

This section presents the results of the elicitation. First, we present the results of the thematic analysis and the subsequent categorization. Then we present the weights and the underlying rationales obtained by instantiating the ISO/IEC 9126 by the DEJ-based and the TA-based approach, respectively. Finally, the results of the comparison and consolidation are summarized.

The Thematic Analysis and Categorization of Terms. The thematic analysis provided categories of non-functional characteristics extracted from coding the AALIANCE Ambient Assisted Living Roadmap. The categories obtained by participant C were: adaptive (18), availability (7), interoperability (23), reliability (13), security (25), usability (14), embedded (1) and cost awareness (3). The categories obtained by participant D were: embedded (4), adaptive (48), available (11), interoperability (25), security (31), user friendly (22), compliance (1). The numbers in the parentheses represent the number of occurrences of the relevant codes. All codes collected were distinctly included in one of the categories. For example, the category *available* obtained by participant D, included the following 11 codes: *always on, access, ease-of-use, available* (3), *connected* (2), *interaction* (2), and *availability of information*. Full traceability to the sources of the codes in the AALIANCE Ambient Assisted Living Roadmap has been documented. Note that there are some discrepancies between the underlying numbers of codes for

Table 1. Results of the elicitation

Quality characteristic	Weights (DEJ)	Weights (TA)	Consolidated weights
Functionality	High	High	High
Reliability	Medium	Medium	Medium
Usability	High	High	High
Efficiency	Low	Low	Low
Maintainability	Medium	Medium	Medium
Portability	High	High	High
Effectiveness	High	High	High
Productivity	High	Low	High
Safety	Medium	High	Medium
Satisfaction	High	Low	High

same categories deduced by C and D. This is due to independent coding and categorization approaches by these two participants.

Weighting of Quality Characteristics. Table 1 presents the results of the instantiation of ISO/IEC 9126 with respect to AAL. The first column specifies the name of the quality characteristic. The first six characteristics are external and internal, while the remaining four ones are quality in use characteristics as defined by ISO/IEC 9126. The second column presents the weights assigned to the quality characteristics during the DEJ-based elicitation (by participants A and B). The third column specifies the weights assigned to the quality characteristics during the TA-based elicitation (by participants C and D). The fourth column lists the consolidated weights (by all four participants).

Rationale for Weights Provided by Participants A and B. While weighting the quality characteristics, participants A and B argued that the AAL services distinguish from software/services in general in the sense that the latter can be assigned requirements by being considered in isolation. The AAL services must, however, function in close interaction with humans, and are often embedded. A and B argued that high importance of functionality characteristic is due to its security sub-characteristic. Privacy and data protection needs are grounded in sensitive health information, personal information and other sensitive information. Interoperability (sub-characteristic of functionality) is important due to the heterogeneous and dynamic nature of the AAL systems, as well as many components having different configurations. Reliability should have medium weight as most AAL services/systems are dedicated to empowerment and quality of living, and are not life sustaining. Usability should be high, since consumers of the AAL services typically are novices with limited experience with the AAL technology. The expectation of minimal interaction is present at both user and care giver side, but this can be changed over time when AAL has become more widespread. Efficiency was weighted low, since AAL services seldom have real-time requirements and are not life sustaining. Some components (e.g. mobile entities and sensors) may, however, require low energy and broadband consumption. Maintainability scored medium due to the need for analyzability and early detection of faults, which is crucial for trust. AAL systems must also be easy to test, update and perform changes on, in addition to being stable during modifications. Portability was argued to be highly important due to the need for installability and co-existence. Additionally, limited technology mastering at the consumer and care giver side, varied application areas of the components, as well as frequent need for adaptation make portability important. Effectiveness (which stresses the functionality) was assigned high importance primarily for the assisted person, and secondarily for the care giver. The high importance of productivity is assigned primarily from the perspective of the formal care giver, provided the objective to improve efficiency of the welfare sector. The medium importance of safety is due to the fact that many AAL services are not safety critical or life sustaining, but rather a supplement to the existing care services. The high importance of satisfaction is due to the voluntary adoption of the AAL services by the assisted person. As such, the services should be positively experienced. For formal care givers, satisfaction is important due to the AAL market domain being rather immature.

Rationale for Weights Provided by Participants C and D. Rationale provided by C and D was solely based on the results of the thematic analysis (based on the relevance of the related categories for the quality characteristics, as well as number of underlying codes for each category). Thus, when providing the rationale, C and D jointly linked the relevant above mentioned categories that have been extracted, to the ISO/IEC 9126 quality characteristics. The links were as follows (where the italic categories stem from the thematic analysis conducted by D and the underlined ones stem from the thematic analysis conducted by C): functionality (*interoperability, security,* interoperability, security), reliability (*security,* reliability), usability (*user-friendly,* usability), efficiency (no categories found), maintainability (*adaptive, interoperability,* adaptive, interoperability), portability (*adaptive, interoperability,* adaptive, interoperability), effectiveness (*user-friendly, adaptive, available,* usability, availability), productivity (no categories found), safety (*security, user-friendly,* security, reliability, usability), satisfaction (no categories found).

Comparison and Consolidation of the Results. The results of the elicitation show almost full agreement between the two approaches. Functionality, usability, portability and effectiveness were evaluated as highest priority quality characteristics in both approaches. Reliability and maintainability were assigned *medium* weight, and efficiency was assigned *low* in both approaches. The only deviation of the results between the two approaches arose with respect to weights of productivity, satisfaction, and safety characteristics. The former two could not be linked to any category in the rationale by the TA-based approach. Moreover, there was in the TA-based approach some uncertainty in the rationale for safety and efficiency. The weights of these four characteristics were by far most uncertain ones in the TA-based elicitation. The reason is that the TA-based approach analyzed only the AALIANCE Ambient Assisted Living Roadmap and extracted solely quality-related aspects. Therefore, the consolidation relied more on the DEJ-based weights of these four characteristics. The TA-based weights of the overall six quality characteristics were considered to be substantiated with higher reliability compared to the weights of productivity, satisfaction, safety, and efficiency. The TA-based weights of those six characteristics were moreover consistent with the DEJ-based ones. Therefore no further consolidation of these was needed. The consolidation has assumed that the DEJ-based weights are relatively reliable due to the DEJ-based weights being founded on the expertise of the participants in the AAL domain, as well as due to the generality of the universAAL reference architecture. Thus, the consolidation of the deviating characteristics concluded that productivity and satisfaction should be assigned *high* weight, while safety was assigned *medium* weight. All ISO/IEC 9126 quality characteristics have been covered. No need for specifying any additional AAL-relevant quality characteristics was discovered in any of the two elicitation approaches.

5   Discussion

The validity [5] of the findings depends to a large extent on how well the threats have been handled. Validity focuses mainly on aspects such as correctness of set-up, quality of process, composition of participants, and accuracy of measurements. Reliability, on the other hand, is concerned with demonstrating that the process can be repeated with the same results. Among the main questions are: did we objectively extract the relevant codes and categories during the thematic analysis; did the participants have the same understanding of the underlying AAL needs; did the participants consistently interpret the quality characteristics; is counting of code occurrences in the thematic analysis fair, since their number is not necessarily proportional to their importance; is the composition of the participants representative; are the AALIANCE Roadmap and the universAAL reference architecture representative as the underlying documentation; are the findings applicable for AAL in general and all instances of AAL?

One of the major threats to validity is the possible bias of the thematic analysis. Both the subjective nature of it and the fact that the number of code occurrences (in the AALIANCE Roadmap) may be a weak indicator of importance, are two obvious threats. The first one was addressed by independent analysis of the two participants (C and D). Nevertheless, we acknowledge that solely counting the number of occurrences obviously imposes a threat to validity. The differences of the categories deduced in the thematic analysis also indicate the subjective nature and partial discrepancies. The second threat was addressed by having both TA-based and the DEJ-based approaches. However, the DEJ-based approach also has uncertainties which at a few occasions could be observed in the form of disagreements between the participants A and B. The discrepancies of the results between the two approaches indicate partial uncertainty of the results in general. Moreover, we have used abstract specifications of the AAL (Roadmap and Reference Architecture). It is uncertain to what degree the set of proposed characteristics (or different subsets thereof) are relevant for specific instances of AAL systems and services. Thus, validation in different instances and using different expert groups is needed. Further validation of the results and evaluation of the approach have therefore to be part of the future work.

Thus, due to the subjective nature of the problem, the conclusions are far from being definitive. Preferably, more experts should have been involved to reduce the bias. Nevertheless, we argue that this paper proposes an initial set of weighted characteristics that should be considered in AAL. In addition, the paper proposes an approach for doing the elicitation in practice by combining two methods (TEJ-based and TA-based). As a part of the validation, the approach can in the future be applied by other researchers in the field. More iterations on the different data sets and by the different communities, may gradually result in more reliable elicitation of quality characteristics.

6   Related Work

Multiple national and European research projects have aligned efforts in the fields of AAL [1,2]. They have, however, mainly concentrated on specification and development of AAL platforms. As a result, well-known AAL platforms have emerged, such as universAAL [12], OASIS [10] and OpenAAL [11]. These platforms address the different quality requirements to a varying degree, without systematically following any specific framework. Therefore, the quality in the respective platforms is difficult to compare, measure and assess. The relevant research initiatives have so far mainly focused on the functional aspects of the AAL systems, without proposing a common ground for quality handling in AAL.

Antonio et al. [4] have evaluated the leading AAL platforms with respect to a set of pre-selected quality characteristics. Their work differs from ours in three manners: 1) the rationale for selection of the quality characteristics is not presented and we do not know to what degree it has been systematic; 2) the quality characteristics selected have been treated as equally important; and 3) they have evaluated fulfillment of the selected quality characteristics in the context of AAL platforms, not AAL systems and services in general.

Walderhaug et al. [16] argue that most of the AAL systems developed address the needs of the end users but have failed to achieve a large market penetration. They argue that this is primarily due to not sufficiently addressing the quality requirements posed by the health care organizations. They present the initial steps towards a framework for measuring quality by building on relevant ISO standards. In their paper, Walderhaug et al. pinpoint selected quality characteristics and argue, based on their experience, for the importance of those characteristics. Our approach differs by making the elicitation systematic through instantiating the ISO/IEC 9126 standard and weighting the quality characteristics relative to each other.

7   Conclusions and Future Work

This paper has been motivated by the lack of explicit specification of the most important non-functional requirements for the AAL systems and services. We propose an approach to elicit, weight and document such requirements in the AAL context. We have experienced that the approach undergone provides valuable knowledge about the AAL in general and its quality requirements in particular. We also provide an initial overview of the important quality characteristics of AAL. These characteristics can be understood as a necessary and adequate complement to existing reference architectures, and serve as guidance for service developers and providers. As such, our results contribute to a common understanding of what AAL is.

Our ongoing and future work includes further validation of the results presented, as well as development of a method for assessing quality fulfillment in AAL. This method is intended to be based on methods for security risk assessment and modeling. In addition, we intend to facilitate use of templates and checklists in order to improve efficiency and correctness of the AAL-specific requirements elicitation.

Acknowledgments. This work has been conducted as a part of the FRISK project funded by the SINTEF Group, as well as a part of the NESSoS (256980) network of excellence funded by the European Commission within the 7th Framework Programme.

References

1. AAL Joint Programme, www.aal-europe.eu (accessed November 18, 2012)
2. AAL Open Association, www.aaloa.org (accessed November 18, 2012)
3. Aarts, E., Marzano, S. (eds.): The New Everyday: Visions of Ambient Intelligence. 010 Publishing (2003)
4. Antonino, P.O., Schneider, D., Hofmann, C., Nakagawa, E.Y.: Evaluation of AAL Platforms According to Architecture-Based Quality Attributes. In: Keyson, D.V., Maher, M.L., Streitz, N., Cheok, A., Augusto, J.C., Wichert, R., Englebienne, G., Aghajan, H., Kröse, B.J.A. (eds.) AmI 2011. LNCS, vol. 7040, pp. 264–274. Springer, Heidelberg (2011)
5. Cook, T.D., Campbell, D.T.: Quasi-Experimentation: Design and Analysis Issues for Field Settings. Houghton Mifflin Company (1979)
6. Horizon 2020 – the framework programme for research and innovation. The European Commission, COM(2011) 808 final (2011)
7. Ezzy, D.: Qualitative analysis: Practice and innovation. Allen & Unwin (2002)
8. Farshchian, B.A., Mikalsen, K.H.M., Reitan, J.: AAL technologies in rehabilitation - Lessons learned from a COPD case study. In: Handbook of Ambient Assisted Living - Technology for Healthcare, Rehabilitation and Well-being. Ambient Intelligence and Smart Environments, vol. 11, pp. 549–566. IOS Press (2012)
9. International Organization for Standardization/International Electrotechnical Commission. ISO/IEC 9126 – Software engineering – Product quality – Part 1-4 (2001-2004)
10. OASIS: Open architecture for Accessible Services Integration and Standardisation, http://oasis-project.eu/ (accessed November 18, 2012)
11. OpenAAL platform, http://www.openaal.org/ (accessed November 18, 2012)
12. Universal Open Platform and Reference Specification for Ambient Assisted Living, www.universaal.org (accessed November 18, 2012)
13. The universAAL reference architecture. universAAL project deliverable D1.3-C (2011)
14. van den Broek, G., Cavallo, F., Wehrmann, C.: Ambient Assisted Living Roadmap. VDI/VDE-IT AALIANCE Office (2009)
15. Web Services Glossary. W3C Working Group Note (February 11, 2004), http://www.w3.org/TR/2004/NOTE-ws-gloss-20040211/ (accessed November 14, 2012)
16. Walderhaug, S., Mikalsen, M., Salvi, D., Svagård, I., Ausen, D., Kofod-Petersen, A.: Towards Quality Assurance of AAL Services. In: Proceedings of the 9th International Conference on Wearable Micro and Nano Technologies for Personalized Health (pHealth 2012). Studies in Health Technology and Informatics, vol. 177, pp. 296–303. IOS Press (2012)

universAAL: Provisioning Platform for AAL Services

Roni Ram[1], Francesco Furfari[3], Michele Girolami[3,4],
Gema Ibañez-Sánchez[5], Juan-Pablo Lázaro-Ramos[6], Christopher Mayer[2],
Barbara Prazak-Aram[2], and Tom Zentek[7]

[1] IBM Research, Haifa, Israel
[2] AIT, Austrian Institute of Technology GmbH, Vienna, Austria
[3] ISTI-CNR, Pisa, Italy
[4] University of Pisa, Pisa, Italy
[5] Universidad Politécnica de Valencia, Valencia, Spain
[6] Soluciones Tecnológicas para la Salud y el Bienestar, S.A. (TSB), Valencia, Spain
[7] FZI Forschungszentrum Informatik, Karlsruhe, Germany

Abstract. universAAL is a European research project that aims at creating an open platform and standards which will make it technically feasible and economically viable to develop Ambient Assisted Living (AAL) solutions. It defined hardware and software infrastructure for smart environments called AAL Spaces, which enable context sharing and reasoning about activities carried out by the assisted person. AAL Services developed with the universAAL platform may be a combination of hardware, software and human resources. Tools for the development, publishing and provisioning of such services have been defined to support the whole chain of stakeholders involved in the AAL domain. The paper focuses on the provisioning of AAL Services by describing the main components involved in the service life cycle.

1 Introduction

One of the problems related to the adoption of AAL Services in real life is the lack of an economically viable platform for the provisioning of AAL Services. Although there were several initiatives funded by the EU in the recent years to create a platform for AAL Services that hides the complexity of the underlying infrastructure [1], none of them was aimed to produce a platform that groups all the needs of the AAL community under the same umbrella.

The goal of the universAAL project [2] is to produce an open platform that facilitates the creation of a thriving market for AAL Services by making new solutions affordable and simple to develop, find, configure, personalize and deploy. The term "AAL Service" refers to the software artefacts, hardware items and human resources that all together compose a useful service for older adults. universAAL's vision is to make the download and setup processes of AAL Services as simple as downloading and installing software applications on a modern operating system.

universAAL's first step towards this objective is to produce a public open platform that works as an intermediate layer between the operating system and the application itself [3]. In order to support the creation of a driven market community for AAL,

A. van Berlo et al. (Eds.): *Ambient Intelligence – Software & Applications,* AISC 219, pp. 105–112.
DOI: 10.1007/978-3-319-00566-9_14   © Springer International Publishing Switzerland 2013

universAAL worked out the uStore, a one-stop-online shop for AAL Services, inspired by the Apple's App Store concept. uStore serves as the ecosystem for all involved AAL stakeholders, namely developers, service providers, research groups as well as the end users and their caregivers. The goal of uStore is to speed up the adoption and spreading of AAL Services by making it easier and more straightforward for the end users to get the services they are looking for.

The main components of the universAAL platform that enable the provisioning process from the developer, through the service provider, to the deployer and the end users are described in Section 2. Section 3 gives an example of an innovative AAL Service that utilizes the smart atrefacts of the universAAL runtime environment and is able to demonstrate the full provision cycle. In section 4 the results of preliminary evaluations that assess the design, usability and quality of the relevant universAAL components are presented. The last section concludes and describes plans for future work.

2   The universAAL Platform

The universAAL platform is built around three pillars: *Runtime support, Development support* and *Community support*. Each of these pillars intends to provide support to different stakeholders. The *Runtime support* enables the cooperation among the different types of devices, sensors and services deployed in the environment in order to assist the person in everyday life activities. It consists of three functional layers: the execution environment, the generic platform services and the AAL platform services. The execution environment extends the native system layer of smart devices and hence hides the distribution of these nodes as well as the possible heterogeneity of their native system layers. It is implemented as a middleware layer facilitating the discovery of devices and services and providing communication capabilities to the upper application layer. Currently, an Android and Java/OSGi version of the middleware are available. The generic platform services are building blocks common to all AmI-based systems; they are used to share context information like user position, user profile and detected activities whereas AAL platform services are domain specific services.

The *Development support* provides tools that facilitate software development based on the universAAL platform and reuse of its components. The main component is an Eclipse based application called AAL Studio. The tools allow the developers to abstract from the actual deployment environment.

Finally, *Community support* is focused on the one-stop-online shop for universAAL services. Service providers and developers can offer their AAL Services in the store while beneficiaries can search and shop complete AAL Services that address their needs. The store is the main meeting point of all involved stakeholders.

The main components that take part in the provisioning cycle are listed below.

The Runtime Environment
The runtime environment allows nodes of a network to cooperate in an environment that is called AAL Space. AAL Spaces are smart environments equipped with well defined software infrastructure [4]. The devices embedded in such environments operate collectively using information and intelligence that is distributed in the

infrastructure. AAL Spaces are classified in profiles, each identifying the typical set of devices used in a specific AAL scenario and the available infrastructure; we distinguish between private space profiles, like homes, versus public space profiles, like supermarkets.

AAL Space actual configurations consist of the nodes (peers) belonging to the space, the distributed applications installed in the space, ontologies dynamically added with the installed applications, and managers which are mandatory and optional building blocks realizing the software infrastructure. One peer of the AAL Space is elected as *coordinator* and it is responsible for the initialization and the announcement of an AAL Space by propagating a space signature called AAL Space Card. Other peers can dynamically join and leave AAL Spaces by exploiting the discovery and peering capabilities of the middleware. In particular, the peering facility allows to exchange security information and to configure the communication channels among the peers. Several communication paradigms such as unicast, reliable multicast and broadcast are available based on pluggable protocol connectors for SLP, SSDP and JGroups communications [5,6,7].

Central component of the provisioning mechanism is the *Deploy Manager* – responsible for managing the installation and un-installation of applications, also called Multi-Part Applications (MPA). A MPA is a distributed application that can be deployed in one single peer or in a number of peers of the AAL Space. The MPA may be composed by a set of application parts; every part wraps one or more software artefacts that are installed into a target peer. Depending on the operating system adopted by the target peer, software artefacts may be very diverse resources like: .exe file, jar file, libraries (.dll or .so), images, and so on. The MPA package (manifest file, licenses and parts) is compressed and sent to the Deploy Manager which is responsible for analyzing the MPA manifest and for delegating the installation of the parts to the target peers by following a deployment schema similar to other remote management specifications [8,9].

The Online Shop

The universAAL one-stop-online shop, uStore, is an eCommerce platform customized to the unique requirements and use cases of the AAL domain. It provides the ability to conduct commerce and specifically assists in publishing, purchasing and deploying AAL Services. Comparing uStore to the most successful solutions that exist today in the market, like Apple's App Store and Amazon, reveals some significant features that do not exist or are not highlighted in the classic eCommerce solutions: *Advanced products* - AAL Service as a composition of software, hardware and human resources. *Service composition* - dynamic composition of AAL Services at purchase time. *Personalization* - based on the end user and the AAL Space profiles. *Accessibility* – better accessibility for uStore main clients; elderly and people with disabilities. *Social commerce* - social interactions among the involved stakeholders and their contributions via blogs, reviews, social networks, etc. *Software management* – management of the software artefacts that compose an AAL Service including versioning support.

Several models were examined for the implementation of the store ranging from the Java content repository API (JSR 170) through content management systems to eCommerce solutions. Ultimately the core of uStore is based on IBM WebSphere Commerce[1],

---

[1] The official web site, http://www-01.ibm.com/software/genservers/commerceproductline/

one of the two world leaders in eCommerce [10]. IBM WebSphere Commerce as a reliable, secure and scalable solution is extended and customized to support the unique aspects of universAAL.

The high level architecture of uStore is composed of three different blocks: *Inner
components* – IBM WebSphere Commerce server (including IBM HTTP server and IBM
DB2) and Sonatype Nexus [11] as the Content Management server for managing the
software artefacts. *Web interfaces* - the Administration Console for operating the store,
the Management Center for managing the uStore catalog by the AAL developers, service
providers and legal authorities and the store site that can be accessed by all the stakeholders. *External interfaces to universAAL tools* - AAL Studio supports developers by simplifying the packaging and deployment of MPA in the uStore and the universAAL Control
Center (uCC), an extendible console running in the AAL Space, helps the end user
(deployer) to select, deploy and install a MPA from the uStore.

The Deployment Tool
The uCC manages the deployment of the software artefacts according to the refer-
ence-deploy-process [12]: *Installation* - this step includes the acceptance of the service license agreement by the end user and the registration of the application in the
uCC database. *Configuration* - this phase is split into the adaptation of the service to
the end users' needs and the collection of customization information for the deployment layout. Based on the application configuration a new view is generated in which
the deployer adds all required properties for the service. The values entered are stored
and used at runtime. The deployment layout is set by the deployer and being used by
the Deploy Manager to distribute application parts to the specified peers in the AAL
Space. *Maintenance* - the last phase provides uCC tools to observe the runtime environment and its services and provide additional functionalities during runtime:
un-installation, managing, reconfiguration, etc.

3   A Practical AAL Service

An AAL Service is the added value perceived by a person. The combination of software, massive hardware distributed in the local space, human resources and processes
enables the person to increase his autonomy and independence level.

Long Term Behavior Analyzer (LTBA) AAL Service
The main goal of the LTBA service is to measure the status of an older adult and to
detect the risk of becoming a frail elderly in order to apply intervention actions that
extend the time the older adults can live autonomously [13]. In the end, this service uses
the sensors distributed in the environments to detect changes in certain parameters related
to the daily habits of an elderly person: functional, social, mental and medical. Those
parameters are directly related to certain common diseases, quality of life and the frailty
level. The information generated by the system is not only raw information but especially
new knowledge coming from the activity detection. It is presented in an understandable
way to professional experts and to relatives who analyze the information in order to help
in the continuous management process of those patients.

This service does not offer much to the assisted person himself. They can access the system by a TabletPC or a similar device to start/stop it (to empower them with control) as well as select which information they want to share with their relatives. However, since this is a preventive service, there is no significant enjoyable feature for them.

LTBA is based on the following technologies: *KISS (Keep it simple stupid) approach* – hide the complexity from all involved users, not only from the older adults. *Sensors* - acquire ambient information about user activities: magnetic contacts in doors, windows and smart appliances, energy smart meters, localization and identification of the user, presence detectors strategically placed, wearable fall detectors, etc. *User context* – user agenda, external information (weather, public events, etc.), user profile, health record, etc. *Artificial intelligence techniques* - as workflow mining and rule based systems to analyze the information acquired and produce a meaningful result. *Games* - use of third party developed games to assess mental status and control potential degeneration in cognitive skills. *Structured questionnaires and feedback* - about elderly's status provided by the relatives.

LTBA Benefits from universAAL Platform

The runtime environment provides many features in order to build Ambient Intelligence applications and specifically provides many of the features required by the LTBA service. The interaction of the service with the underlying platform is achieved by a group of libraries that the service interfaces with. LTBA is configured with multiple nodes in the AAL Space running instances of universAAL runtime and sharing information among them. The features offered by the runtime platform for the benefit of LTBA are:

- Abstraction: LTBA is able to work with any hardware (e.g. KXN, ZWave and ZigBee devices [14]). By default, universAAL supports the above technologies, though developers can add translators to other technologies/protocols.
- Intelligence: universAAL is providing by default a Drools [15] based reasoner which based on complex rules creates events that indicate new knowledge derived from a rule match. This is a very useful approach to detect activities from a composition of different events with a certain timing pattern.
- Authenticated Remote Access: both technicians and relatives of the assisted person are able to enter the system at home by using a web based interface.
- Reuse of heterogeneous applications: questionnaires are usually legacy applications developed by healthcare experts that can be integrated in the AAL space and easily deployed on some peer by using a multi-part application schema.
- Interaction with Third Party Applications: third party applications can listen to triggered rules and to raw events coming from sensors or ask about current activity being done by the assisted person by using several protocols.

The universAAL platform as a whole provides the full marketing chain for LTBA: development, publishing, purchasing, deployment and utilization (see Fig. 1).

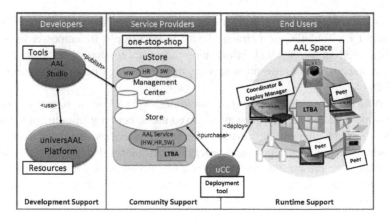

Fig. 1. Long Term Behaviour Analyzer full provisioning cycle

Through uStore developers and service providers can offer the LTBA service and make it available to users under different purchasing models (free, fixed, per use, per time, etc.). For the end users (older adults and their care givers), uStore provides a simple way to find and acquire the service. By acquiring the LTBA service from uStore, required software (applications and device drivers) will be easily deployed to the user's local environment by the uCC tool. New hardware will be ordered, mailed and locally installed. Ideally, the installation process is being done by the end users or by their relatives. In some cases when the installation is more complex (typically involving hardware installation), agreements will be made with the service provider to reserve required human resources both for the deployment and the operation of the service. Finally, the LTBA service is being utilized, providing ongoing measurements on the status of the older adult.

4   Evaluations

In the course of the universAAL project various evaluations and analyses have been performed. The evaluations are ongoing and accompanying activities of the development process. Some results and insights concerning the above mentioned components are summarized in the following.

The definition of the LTBA service has been evaluated in a small scale explorative study with 7 informal/formal caregivers and 4 older adults by means of face-to-face interviews. In general the participants rated the service as interesting and useful. Especially, the prediction of chronic diseases and the generation of easily to interpret daily reports, have been mentioned positively. Nevertheless, there are still some barriers to be solved. The most important one is the invasion of privacy. Older adults do not reject the data collection and analysis per se, but want to be fully informed about the functionality of the service and especially the usage of the collected sensitive data. They want to feel confident with the system and to be able to control it. Other important open questions are related to the cost of service.

universAAL uses scenarios, which are short and non-technical descriptions, as a design tool to get a better understanding, identification and validation of the user needs. These scenarios describe different functionalities of universAAL from the point of view of a certain key stakeholder. For this paper the view of the older adult has been picked since it covers the online uStore as a single access point to services offered by different vendors. The uStore scenarios have been evaluated in an online survey with 10 participants. The main response has been that the concept of uStore is liked by most people, especially as a place to find and learn more about services that can help older adults in their daily life. Nevertheless, people are still skeptical regarding technology, since it is still perceived as something that can isolate assisted persons from personal relationships and that may invade their privacy. It is important to use technology to support and not to replace people, but it will take time to overcome technophobia of older adults.

5   Conclusions

In this article we introduced the universAAL integrated system that provides not only an open and scalable technological platform that facilitates the execution of a broad range of AAL Services, enabling ecosystems of interoperable hardware and software products. Also, provides tools for developing and reusing code for innovative AAL Services, meeting all stakeholders together in a one-stop-online shop and giving support to the end users to manage their AAL Spaces.

universAAL's achievements contribute to increase the AAL market fostering the wider use of a common open platform, tools and established standards. Functionalities as privacy, confidentiality, data protection and an extensive user community building will help to achieve the expected impact of universAAL in the AAL market. Furthermore, parallel activities such as AALOA [16] will ensure the continuity and maintenance of universAAL after the project ends. Further projects funded by the EU, such as ReAAL, will pursue the validation and wide adoption of the universAAL platform while measuring the related socio-economical impact.

Acknowledgments. The research leading to this work has received funding from the EU 7th Framework Programme under grant agreement no. 247950 (universAAL).

References

1. Hanke, S., Mayer, C., Hoeftberger, O., Boos, H., Wichert, R., Tazari, M.-R., Wolf, P., Furfari, F.: universAAL – an open and consolidated AAL platform. In: Wichert, R., Eberhardt, B. (eds.) Ambient Assisted Living. LNCS, vol. 63, pp. 127–140. Springer, Heidelberg (2011)
2. Hanke, S., Eisenberg, V., Tazari, S., Furfari, F., Mosmondor, M., Peres, Y., Wichert, R.: universAAL – An open Platform and Reference Specification for Ambient Assisted Living, Med-e-Tel 2011 Conference Proceedings (2011)
3. Hanke, S., Mayer, C., Kropf, J., Hochgatterer, A.: A Need for an Interoperable Open Source Middleware for Ambient Assisted Living - A Position Paper. In: roceedings HEALTHINF2010 Valencia - Special session on Open Source in European Health Care (2010)

4. Tazari, M., Furfari, F., Fides Valero, Á., Hanke, S., Höftberger, O., Kehagias, D., Mosmondor, M., Wichert, R., Wolf, P.: The universAAL Reference Model for AAL. In: Handbook of Ambient Assisted Living, vol. 11, pp. 610–625. IOS Press (2012)
5. Edwards, W.K.: Discovery systems in ubiquitous computing. IEEE Pervasive Computing 5(2), 70–77 (2006)
6. Zhu, F., Mutka, M.W., Ni, L.M.: Service discovery in pervasive computing environments. IEEE Pervasive Computing 4(4), 81–90 (2005)
7. Sales, L., Teofilo, H., D'Orleans, J., Mendonca, N.C., Barbosa, R., Trinta, F.: Performance Impact Analysis of Two Generic Group Communication APIs. In: IEEE International Conference on Computer Software and Applications, vol. 2, pp. 148–153 (2009)
8. UPnP Device Management Specification, UPnP Forum (February 16, 2012)
9. Hillen, B.A.G., Passchier, I., van Schoonhoven, B.H.A., den Hartog, F.T.H.: Remote Management of Non-TR-069 UPnP End-User Devices in a Private Network. In: IEEE Conference on Consumer Communications and Networking, CCNC 2009 (2009)
10. Alvarez, G., Clark, B.: Top Trends in E-Commerce and the 2010 E-Commerce Magic Quadrant, slide 9, http://www.soneti.net/Ficheiros/Gartner.pdf
11. Andersen, T.J., Amdor, L.E.: Leveraging Maven 2 for Agility. In: Agile Conference 2009, AGILE 2009, pp. 383–386 (2009)
12. Stumptner, M.: An overview of knowledge-based configuration. AI Communications 10(2) (June 1997)
13. Rogers, R., Peres, Y., Müller, W.: Living longer independently – a healthcare interoperability perspective. e & I Elektrotechnik and Informationstechnik 127 (August 2010)
14. Baronti, P., Pillai, P., Chook, V., Chessa, S., Gotta, A., Hu, Y.F.: Wireless sensor networks: a survey on the state of the art and the 802.15.4 and zigBee standards. Computer Communications 30, 1655–1695 (2007)
15. Drools rule engine, http://downloads.jboss.com/drools/docs/4.0.7.19894.GA/html/index.html
16. Furfari F., Potortì F., Chessa S., Tazari M., Hellenschmidt M., Wichert R., Gorman J., Kung A.: The AAL Open Association Manifesto. In: Lukowicz, P., Kunze, K., Kortuem, G. (eds.) EuroSSC 2010. LNCS, vol. 6446. Springer, Heidelberg (2010)

Separating the Content from the Presentation in AAL: The universAAL UI Framework and the Swing UI Handler

Alejandro Martín Medrano Gil[1], Dario Salvi[1], María Teresa Arredondo Waldmeyer[1],
Patricia Abril Jimenez[1], and Andrej Grguric[2]

[1] Life Supporting Technologies, Universidad Politécnica de Madrid, Madrid, Spain
{amedrano,dsalvi,pabril,mta}@lst.tfo.upm.es
[2] Research and Innovations Unit, Ericsson Nikola Tesla, Krapinska 45, 10000 Zagreb, Croatia
andrej.grguric@ericsson.com

Abstract. When it comes to user interaction elderly and people with special needs have special requirements, that evolve over time. Ambient Assisted Living (AAL) envisions ubiquitous, intuitive human-machine interaction running on heterogeneous devices (tablets, televisions, smart-phones, etc.) and by means of multiple modalities (graphics, voice, gesture recognition, etc.). This vision often represents a strong technological challenge for any developer of AAL services. In order to relieve the developer from the burden of repeatedly adapting the interfaces, the universAAL platform [6] offers a framework for user interaction that separates the content, that is exchanged between the user and the application, and its actual representation. The main drawback of this solution is that it delegates all the responsibility of the representation of the content to the platform, which, in turn, must be able to adapt to all kinds of showable messages. This paper explains how a so-called *UI Handler* has been designed and implemented to render messages in universAAL, as well as the results of its technical and usability evaluations.

Introduction

The European population is becoming increasingly older. As a result of this demographic trend and the global economical crisis, enormous effects on expenditures in welfare and healthcare are putting the sustainability of European social systems in danger. The AAL paradigm faces this challenge by envisioning the utilization of Information and Communication Technologies (ICT) in support of elderly population, with individuals surrounded by networked and computing technologies, unobtrusively embedded in their environment, helping them to better manage their daily lives.

One of the biggest obstacles to its success is the distance between technologies and the real needs of the population. When it comes to user interaction (UI) this is particularly important because AAL technologies should serve people with special characteristics like reduced mobility, visual impairments and lack of familiarity with computers [7]. AAL aims at a natural, simple, effortless, device-independent, and reactive interaction with users; allowing continuous, autonomous and ubiquitous service adaptation with the aim of engaging and empowering users. Several interaction channels

A. van Berlo et al. (Eds.): *Ambient Intelligence – Software & Applications*, AISC 219, pp. 113–120.
DOI: 10.1007/978-3-319-00566-9_15   © Springer International Publishing Switzerland 2013

are available nowadays for interacting with users, different devices (televisions, smart-phones, tablets etc.) as well as alternative communication means (e.g. graphics, sound, voice, gestures etc.), and AAL applications should make use of as many of these means as possible in order to accommodate the strong and evolving needs of older adults. In order to abstract technologies, *platforms* are being created, but, although several have already been proposed, none reached a sufficiently wide adoption.

UniversAAL[1] is a European co-funded project that aims at developing the reference open platform "for providing a standardized approach for making it technically fea-sible and economically viable to develop AAL solutions". The universAAL platform includes a framework for user interaction for abstracting UI. The framework models the messages that applications can send to the users by means of a semantic model that organizes the content into concepts like dialogues, submits and input/output fields that carry no information about the way they should be represented. The role of delivering messages, as well as collecting feedback from the user, is delegated to *UI Handlers*. Several UI handlers can co-exist in a single set-up, each one dealing with a different device or a specific modality (e.g. a UI Handler for graphics, another for voice, etc.).

Building a UI Handler is daunting task, given that it has to be flexible enough to correctly represent all kinds of messages, regardless of their complexity and, at the same time, be usable and attractive. The reference UI Handler for standard computer graphics included in universAAL, the Swing UI Handler, is the objective of the work explained hereby. The following sections show how the Swing UI Handler has been designed, implemented and evaluated.

1   Methods

1.1   The universAAL UI Framework

The universAAL platform is based on the concept of "buses". A bus is a software arte-fact that abstracts the distribution of the networked nodes sharing a common "middle-ware". Buses can be either event based (with providers and subscribers), or call based (typically with callers and callees). There are currently three buses in universAAL, one dedicated to contextual information, the Context bus, one dedicated to services, the Service bus, and one dedicated to user interaction, the UI bus.

The UI bus is part of the UI framework, which consists of the following components:

UI Bus   provides distributed methods for accessing universAAL instances in relation to user interaction.

UI Model   contains ontological representations of the most common user in-teraction components i.e. the building blocks for sending messages (UIRequests) and retrieving feedbacks (UIResponses) that are shared among the bus peers. It enables modality-neutral description of user interfaces, inspired by the standard W3C XForms.

UI Caller   is the representative of an application, it sends UIRequests to the bus, and manages UIResponses from the user.

---

[1] http://www.universaal.org/

Dialog Manager is a single instance in the AAL space (the connected set of universAAL nodes), that is used by the bus in order to manage system-wide
coordination of each user's dialogs through dialog priority and adaptations.

UI Handlers the components that represent UIRequests to the user and packages
UIResponses to be delivered back to UICallers.

Applications define what they want to present to the user (and what they need from
him/her) in terms of dialogues using previously mentioned UI Model. Once the dialogue
is built, it is encapsulated in a UIRequest that is sent through the UI Bus. The Dialog
Manager then approves and adapts the UIRequest, and the bus routes the message to
the best UIHandler (eg. the physically nearest to the destined user or the most suitable
for the user's needs). UI Handlers unwrap the dialogue coming from the applications
and present it to the user in an appropriate way (through the channel the UI Handler
controls). Once the user provides the required input, this information is injected into
the dialogue and sent back to the application via a UIResponse.

1.2 The Swing UI Handler Internal Architecture

The Swing UI Handler is the default UI Handler of the universAAL platform for standard computer graphics interaction. It is designed to work on standard computers with
a Java SE virtual machine and is based on the Java Swing [8] graphic library.

The design of the Swing UI Handler is based on the Hierarchical Model View Controller Architecture [4]. This architecture decomposes each UIRequest into its building
blocks (ie. the individual UI Model components) and passes them to a set of modules
specialized to render each block. The actual visualization of a block is then delegated to
two layers, as shown in figure 1. The first layer, namely the model layer, is responsible
for creating an instance of a Swing component for each processed block (e.g. a submit is
mapped to a button), these Swing components are generated with default look and feel
and filled with the content provided with the request (e.g. the title of the button). The
second layer, namely the LAF layer, is a collection of classes that extend the first layer,
providing the Swing components with the visual details and behaviour (e.g. size, color
and layout), implementing, therefore, the actual look and feel (LAF) of the Handler. A
collection of these "graphical enhancers" is called LAF package.

A LAF, in this context, is different from the standard Java Swing "look and feel".
While the swing proposal focuses on system-wide visual customizations; our proposal
is block centred customizing each block independently according to its context. This
differentiation enables first, for LAF packages, to use the swing "look and feel" methodology to ease the implementation of system-wide features; and second, for LAFs to be
designed for specific impairments. As an example of both possibilities a LAF package
can use swing's "look and feel" mechanism to set default visualizations like background
colour or font size for all swing components, but the block centred proposal enables for
it to adjust size and colour so the information is better differentiated by the user, even if
rendered with the same swing component.

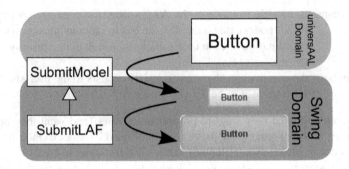

Fig. 1. Example of the model and LAF layer interaction for the case of a submit. The model layer is based on the universAAL UI framework model, and provides an instance of a Swing button, while the LAF layer extends the model providing a visually enriched instance of the button.

In order to allow developers to create LAFs adapted to certain accessibility challenges, an API is offered to initialize LAF packages, configure global settings, and react to users' logins/logoffs so user-specific customizations can be reflected. The power of this architecture resides on the capability of the Swing UI Handler to use different LAF packages, which can be specialized for different UI challenges, while maintaining the development of these LAFs packages easy and content management free.

1.3 The UI Development Guidelines

With the aim to facilitate the work of developing effective and easy to use applications, the Swing UI Handler comes with a manual containing a set of good practices for developing applications' UIs and/or LAF extensions. These good practices cope with two main challenges. On one hand they recommend typical usability patterns for elderly people, providing suggestions for developing intuitive interfaces that minimize the cognitive effort for the users. On the other hand, they provide a methodology that guides the application developer into the implementation of the interfaces using the universAAL UI framework.

The guidelines comprise a combination of the principles stated in the User Centred Design, a set of relevant standards [1,2] and suggestions from experts in the field [3,5,7]. The proposed approach integrates the classic user-centred iterative methodology, from requirements assessment to user validations with tasks specific for development with the universAAL platform (e.g. suggestions about colours and metaphors to be adopted in LAFs, a scheme for the organization of the dialogues etc.).

2 Evaluation Results

The Swing UI Handler has been evaluated under both the developer's perspective and the user's perspective.

2.1   Technical Evaluation

The evaluation with developers involved two programmers involved in the project who developed a set of LAF packages and some supporting applications for testing the outcome of the LAFs. One default LAF was already provided together with the handler while three other LAFs were produced by the participants of the evaluation (see figure 2).

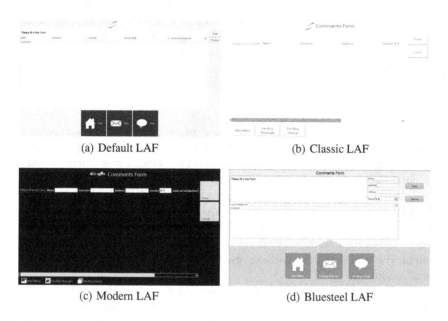

(a) Default LAF   (b) Classic LAF

(c) Modern LAF   (d) Bluesteel LAF

Fig. 2. Samples of LAF packages showing the same application's form. (a) is the default package, (b) and (c) were developed by one participant and (d) by another.

After the development had concluded, the programmers were asked to answer a questionnaire. Four aspects were evaluated: 1) if the architecture and design are easy to understand, 2) whether the documentation provided is enough and of sufficient quality, 3) how easy is to develop custom LAF packages, and 4) the potential to extend the design and/or the implementation to other functionalities and technologies.

The answers show that the time required to develop a fully customized LAF package is around 2 weeks, including the training. The learning curve is gentle, but can still be improved for instance by allowing simple changes in the configuration of the default LAF. Participants also stated that their preferred source of knowledge and skill acquisition were the examples. This implies that there must be well defined LAF examples and templates where the programming patterns are clearly shown, and these samples must be well documented.

The most critical detected point is the availability of layouts for organizing the content. In the proposed implementation, layouts are left as part of the LAF development, but the fact that the Swing framework has a small set of layout managers constrains the developer's options. With respect to the current implementation, a richer collection of layouts is needed, or preferably a component that would automatically select the best-fitting layout on the basis of the content to be shown.

A secondary detected issue is related to the capability of overlaying dialogs (e.g. showing pop-ups) which is responsibility of a component that is already separated within the handler's architecture, and activated through the handler's configuration parameters.

Finally, the participants believe that the current architectural design is easily portable to other UI technologies, even for other modalities or devices. Porting the paradigm to other technologies implies the implementation of a new UI Handler, and therefore development of a model for the selected technology, then LAFs have to extend that model. Some candidate technologies mentioned by participants are Android for mobile devices, and GWT and other html-based technologies for web-based modality.

2.2   Usability Evaluation

For what regards the evaluation from the users' perspective, 8 usability experts were involved to assess the graphical outcomes of the Swing UI Handler. Seven engineers and one psychologist, with more than 5 years experience in usability with elderly population from four different countries participated.

The assessment required the preparation of a set of sample applications that showed some classical patterns of interfaces that an AAL application can implement. Two scenarios were created, the first scenario included three sample applications with simple interfaces and navigation flows, while the second scenario included eleven sample applications with very complex interfaces, used to test the behaviour of the handler under extreme conditions.

The actual evaluation consisted of two phases. In a first phase, experts had to run the sample applications and perform simple, repeatable operations, called walk-throughs [9], analyse the presence of usability issues and provide suggestions for improvements. Walk-throughs were designed in order that evaluators had the chance to see all the main functionalities of the handler, and how typical dialogues were organized on the screen. In a second phase experts had to fill in a questionnaire examining whether the handler met a set of standard usability heuristics as defined by Nielsen in [10]. Both simple and complex scenarios were evaluated separately and the most advanced LAF (namely Bluesteel pictured in figure 2(d)) was used. Some alternative implementations of some graphical details were also assessed separately.

The results of the evaluation show that on average the Handler produces outputs that are suitable for elderly users if the rendered dialogue is simple enough, while, with more complex examples, the implemented layout fails to organize the information in an efficient way. Participants found the look and feel of the graphical elements coherent and attractive. Experts also appreciated the way on-line help is used to guide the users towards the dialogues and suggested specific improvements like a generic "back" button for cancelling operations and undo-redo commands for recovering from errors while filling the forms. Also the navigation through running applications was assessed with different options, and the preference of the evaluators fell on using a general "home" menu where all applications are listed both for starting them and resuming their usage when in background.

3   Conclusions

The universAAL UI framework is based on the separation between content and actual representation of user interfaces in a highly distributed environment. Although beneficial for applications development, this separation raises big challenges in the implementation of the components responsible for the visualization of the interfaces, the UI Handlers. The Swing UI Handler proves that it is certainly feasible to design and implement such components, but it is not a simple task. One important aspect to be considered is the suitability of its internal architecture and the separation between models and look and feel.

The preliminary evaluation conducted on the Swing UI Handler shows that the current implementation still needs many improvements, but also that good design decisions were taken and that the work is promising. Concretely the technical evaluation shows that the internal design of the Handler is well understood and that LAF package extensions are simple to implement. Nonetheless, the usability evaluation proved that the major point of failure is represented by the adopted LAF package, and that the one which was evaluated still has many bugs to be solved.

There are indications that the design principles of the handler can be adopted in other UI technologies, even including other modalities, and the principles of the UI framework and the Swing UI Handler can be also extended to other AAL platforms or domains.

For future development, the results of the evaluation will be taken into consideration for improving the artefact. A second round of evaluations, with more realistic applications and real final users, will be performed. The evaluation will have to assess if real users' needs and preferences are covered and, if these assessments are positive, the results feed a new version of the UI development guidelines.

Acknowledgement. We would like to thank the whole universAAL Consortium for their valuable contribution for the realization of this work. This work has been partially funded by European Union under the 7th Framework Program through the universAAL (FP7-247950) research project.

References

1. ISO/IEC 9241-14: 1998 (E) Ergonomic requirements for office work with visual display terminals (VDT)s. Tech. rep., International Standardation Organisation (ISO) (1998)
2. ISO/IEC 13407: 1999 Human-centred design processes for interactive systems. European Standard EN ISO 13407:1999. Tech. rep., International Standardation Organisation (ISO) (1999)
3. 7.5, S., Zwick, C., Schmitz, B.: Designing for Small Screens: Mobile Phones, Smart Phones, PDAs, Pocket PCs, Navigation Systems, MP3 Players, Game Consoles, 1 edn. AVA Publishing (2006)
4. Medrano Gil, A., Salvi, D., Arredondo Waldmeyer, M., Abril Jimenez, P., Fides Valero, A.: Use of hierarchical model-view-controller architecture for user interaction in aal environments. In: III Woekshop on Technology for Healthcare and Health Lifestyle, ITACA (2011)
5. Cooper, A., Reimann, R., Cronin, D.: About face 3: the essentials of interaction design. Wiley (2012)

6. Hanke, S., Mayer, C., Hoeftberger, O., Boos, H., Wichert, R., Tazari, M.R., Wolf, P., Furfari, F.: universAAL - An Open and Consolidated AAL Platform. Springer, Dordrecht (2011)
7. Hawthorn, D.: Possible implications of aging for interface designers. Interacting with Computers 12(5), 507–528 (2000)
8. Loy, M., Eckstein, R., Wood, D., Elliott, J., Cole, B.: Java Swing. O'Reilly Media (2012)
9. Nielsen, J., Mack, R.L.: Usability Inspection Methods. Wiley (1994)
10. Nielsen, J., Molich, R.: Heuristic evaluation of user interfaces. In: Proceedings of the SIGCHI Conference on Human Factors in Computing Systems, CHI 1990, pp. 249–256. ACM, New York (1990), doi:10.1145/97243.97281

Performance Considerations
in Ontology Based Ambient Intelligence Architectures

Martin Peters[1], Christopher Brink[1], Sabine Sachweh[1], and Albert Zündorf[2]

[1] University of Applied Sciences Dortmund, Germany, Department of Computer Science
{martin.peters,christopher.brink,sabine.sachweh}@fh-dortmund.de
[2] University of Kassel, Germany, Software Engineering Research Group,
Department of Computer Science and Electrical Engineering
zuendorf@cs.uni-kassel.de

Abstract. One limitation that still exists for the use of ontologies in pervasive and ambient intelligence environments is the performance of the reasoning task, which can slow down the use of an application and make a solution inappropriate for some scenarios. In this paper we first present the results of a user evaluation that substantiates the amount of time, that is acceptable (from the point of view of a user) as a delay resulting from the reasoning process in ontology based scenarios. Based on this results we introduce an experimental setup to test the performance of an ontology based architecture. This test shall demonstrate the performance of the state of the art technology without specific performance optimizations and provide concrete measurements for such a setup.

Keywords: Ambient intelligence architecture, ontology, reasoning, performance.

1 Introduction

Pervasive computing, ambient intelligence and smart environments have recently gained a high popularity resulting in a huge number of research projects, e.g. [1] [2]. For the representation of the domain knowledge as well as the context data the use of ontologies is a common way in the field of ambient intelligence and related areas like ambient assisted living.

Ontologies offer a high expressiveness, can be exchanged between different systems and therefore can be reused. They also provide reasoning capabilities to generate new knowledge based on defined rules. Thus, ontologies are often used to define the domain knowledge as well as to build the "intelligence" within an application and to analyze the context information [3].

While smart environments, weather they are based on ontologies or not, may compute complex states to derive further knowledge, they are also used for simple tasks like switching a lamp on or off when the associated light switch was pressed. This is a straight forward task in a static environment where the connection between the light switch and the corresponding light is defined within a program. In a less static environment which is able to dynamically change its behavior (for example switch a different light when the light switch is pressed) and which can be extended through further components at runtime, other mechanisms than defining the behavior in static programs are necessary.

A. van Berlo et al. (Eds.): *Ambient Intelligence – Software & Applications*, AISC 219, pp. 121–128.
DOI: 10.1007/978-3-319-00566-9_16   © Springer International Publishing Switzerland 2013

To accomplish this task, rule engines provide a dynamic and flexible solution which also allow to add and modify rules at runtime. They can be differentiated to semantic reasoners that work on ontological data to compute additional knowledge and business rule engines which trigger actions based on a given fact basis. A semantic reasoner is usually build to provide support for one or more ontology specifications like the Web Ontology Language (OWL). In addition many of them allow to additionally apply user defined rules based on a rule syntax like the semantic web rule language (SWRL) [4] or Jena[1] rules. These rules in turn may be used to define a specific behavior within the form similar to an event-condition-action rule.

Nevertheless, the flexibility gained through the use of rule engines also consumes much more computational power than the execution of static code and thus is often much more time intensive. The resulting delay of a defined action may not be critical in some cases, but in other cases like within the light switch example a few seconds of delay are undesirable.

In this paper we are taking a closer look at the aforementioned problem of delays of expected reactions resulting by the overhead of an ambient environment. After providing an overview of the related work in section 2 we introduce an user study that was performed to approximate the amount of time which is acceptable as a delay for a reaction on a users input in section 3. In this premises we evaluate the time that it takes in the ontology based act-mobile architecture [5] [6] to react on an event or a change in the environment. After discussing some architectural-specific aspects that may influence the use of existing approaches for a better performance, we finally summarize the results and give an overview of future work before we conclude the paper.

2   Related Work

The importance of ontologies in the field of ambient intelligence and pervasive computing architectures can be illustrated by the number of research projects using semantic technologies, like [7] [8] [9] [10] [11]. In [11] the authors present a service oriented architecture that is based on semantic technologies and dedicated to smart homes. In the described work the authors use a context model based on OWL that describes a smart home as well as objects in the smart home. In addition SWRL is used to define rules to create high-level contextual data as well as to create action rules for example to switch a light. The authors mention in their conclusion, that the inference time slows down the use of ontologies without mentioning a specific factor. The architecture proposed in [10] uses a similar approach. The authors use OWL for context representations and SWRL to define inference rules. In [12] a performance evaluation of an ontology based context-aware middleware for smart spaces is done. The proposed architecture is based on agent-technology and uses OWL for context modeling. The reasoning is done using a probabilistic approach based on a Bayesian network. While this approach offers a way to handle uncertain context, the results of the evaluation show that it is only suitable for non time critical applications.

To achieve a higher performance, Agostini et al. proposed a hybrid reasoning mechanism within their CARE [9] middleware. They are keeping the TBox data - the

---

[1] Jena is a java framework for building semantic web applications.

ontological part that holds the definition of concepts and relations between those - static to be able to perform the TBox classification in advance, before the actual information is requested through a context sensitive service. The same concept is partly applied to the ABox data which holds the individuals of the ontology. Instances, that are known before the contextual information is evaluated are used to prepare the ontological model to provide a faster response time on a service request. While the results of the performance evaluation show that this approach still needs seconds for reasoning and rule execution on ontologies with 500 classes and 2000 instances, it scales better than other approaches with respect to the ontology size.

Another hybrid approach is proposed in [13], where the context model gets divided into a static part which is defined by an ontology and a dynamic part which is stored within a relational database. This approach addresses the time consuming reasoning process resulting from frequent changes on the ontology. Thus, selected concepts from the ontology are transferred to the database and both models are connected through so called knowledge connectors. The results of the evaluation show, that the hybrid model performs faster than a pure database model or an ontological model.

Besides the use of hybrid approaches to achieve a higher performance, the use of clustered computers for a distributed and parallel reasoning is considered in several other works like [14] [15]. While these approaches seem to be promising in providing a way to reason about millions of triples in the context of the semantic web, the overhead of distributing the computation on different hosts is too high for smart environment scenarios with just a few thousand triples.

3   Time for Reaction

While the significance of ontology based architectures was already shown in section 2, the following user study shall answer the question, which delay of an action is acceptable for users in a smart environment, to be able to point out an assessment about the suitability of semantic technologies for ambient scenarios.

The experiment shall reflect typical situations in a smart home like switching a light. Accordingly the first test we are doing is letting people press a light switch. The light switch causes a random artificial delay which simulates the reasoning and computation time and finally switches a high power LED-lamp. After the light is on, the user has to decide if the delay between pressing the switch and the reaction on the lamp was not noticeable (answer *a)*), was noticeable but did not bother (answer *b)*), was noticeable and bothered a little (answer *c)*) or was noticeable and bothered a lot (answer *d)*). This process is repeated 55 times with different delays between 0 and 4000 milliseconds, while the delays from 0 to 1000 ms are more fine-graded to get a detailed impression about which delay gets recognized at all.

A similar test-case like described before is used in a second evaluation, where the same setup is used to control a music player. In this case, the laptop itself is the music player, so that there is not additional delay in terms of latency like with the lamp. The user can click the play button and the program will play a song after a random delay. To not influence the user through visual effects on the user interface, the program is multithreaded so that there is no optical relation between the button click and time of delay until the music plays. In addition the song does not start to play from the beginning, so that there will be no silent parts which mistakenly could be identified as a delay by the user.

The tests where performed by 20 users (scientific staff and computer science students) and each of them used both test cases, the lamp and the music player. We have chosen these users with respect to their technical background and their affinity to technical improvements. These users usually have a particularly high demand on technology integrated in their environment. In figure 1 the averaged answers are plotted on the y-axis (from *a)*, was not noticeable, to *d)*, was noticeable and bothered a lot) and the actually delay on the x-axis.

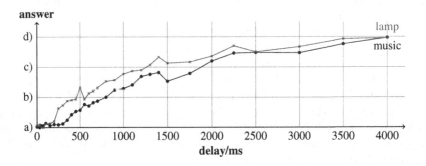

Fig. 1. Results of the evaluation

It can be seen, that there is only a little difference between the lamp and the music player. This difference may be caused through the natural latency the lamp has in terms of communication and computation on the lamp-controller. However, many users described that they were more patiently with the music because they were used to an internet radio which also takes some time to start to play. It can also be seen, that small delays of up to 300 ms nearly do not get recognized by a user while delays from 500 ms to 1000 ms definitely get recognized, but do not bother most of the user in such a smart home scenario. This changes when it comes to delays of up to 1300 ms, which seems to be a limit where the user gets bothered a little. The next limit according to our tests is located at 2200 ms to 2300 ms, which is a delay that really seems to bother. At this point, also the recognized difference between the delay of the lamp and the music gets smaller because the artificial delay dominates the actual delay of the lamp.

4   Experimental Setup and Evaluation

We chose the act-mobile [5] [6] architecture as a reference architecture because of its special characteristics and the possibility, to load an arbitrary owl-ontology. The act-mobile architecture is domain independent and can serve many different connected environments, using different ontologies, at the same time through a client / server architecture. Each environment is connected through a local gateway, which transfers data between the act-mobile server and the local sensors and actuators. The architecture allows to simply load an adequate ontology for a given scenario like for a smart home, but also for a different domain like for a small sewage treatment system. This also implies, that the architecture needs to be scaleable due to many possible requests for different environments and thus for different ontologies.

While the architecture itself is build on java enterprise technologies, the semantic part is build using the Apache Jena Framework. In our setup we use the OWL reasoner provided by Jena in addition to the general purpose rule engine to apply rules for semantic enrichment as well as to react on events. In contrast to most of the aforementioned architectures, the general purpose rule engine does not support SWRL rules but has its own rule language, Jena rules. The rules as well as information about an environment are stored in a database which is independent of the ontology datastore.

There are several things which should be considered in a performance evaluation. On the one hand, we want to measure the time that it takes for one reasoning process to finish. This time may depend on the size of the ontology as well as on the number of rules that are applied. Furthermore this step is independent of other characteristics of the chosen act-mobile architecture and depends on the performance of the chosen reasoner. On the other hand, we want to know how long the overall process takes from pushing a button to a recognizable effect for example on a lamp. Accordingly we are going to investigate the performance on different ontologies with different sizes. The rules that are applied to the ontologies are not only used to provide a higher level of context information (for example, a temperature sensor with a value of above 70°C is a hot sensor) but are also used as action rules for example to switch a light on or off. For the tests the act-mobile architecture was deployed on a server with a 2.4 GHz four core Intel Xeon processor and 6GB of memory.

To define a smart building environment we used the BOnSAI [16] ontology, which is a smart building ontology for ambient intelligence and contains among other things concepts to define a building with floors and rooms as well as sensors and actuators. Based on this ontology we created a model of the building of our department of computer science in Dortmund, containing 10 floors and 110 rooms and lecture halls. Each room in the model was equipped with things like a light, a light switch, an air conditioner and different kinds of sensors. Finally the ontology had a size of more than 9500 triples.

For a second and smaller use case we defined an ontology for small sewage treatment systems and created an environment which finally consisted of about 630 triples. The user-defined rules that were applied to the environments had a varying complexity and partly were based on each other. For example in the small sewage treatment example we created a rule which defines a container to be empty while a second rule fires a notification when a container is empty and some other criteria are met.

4.1   Results

To get an overview of the actually time that it takes from pressing a button until an action is triggered, we simulated a gateway and thus a connected environment that sends messages to the server and expected a specific answer in form of a JMS message that was triggered by a rule. In addition we simply updated values like provided by temperature sensors, which for example could result in controlling the air condition, but we also simulated events. This steps were repeated with different sensors and actuators over 50 times on each environment to get the averaged values which are listed in table 1.

Table 1. Performance evaluation results

	smart building	sewage treatment
preparing time	619 ms	56 ms
reasoning time	3153 ms	92 ms
server process time	3772 ms	148 ms
reaction time	2473 ms	140 ms

The preparing time in table 1 describes the time that was necessary to load the model, the rules and to update or insert the transmitted value within the model. The reasoning time is the amount of time, that the reasoner needed to build the OWL ontology and to apply the given rules. The server process time is calculated by adding the preparing time to the reasoning time. The reaction time in turn is the time measured by the client from sending a JMS message until the expected result in form of a reaction triggered by a rule was returned.

What should be noted is that the average time until an action is triggered and transferred back to a gateway is smaller than the average time the whole update and reasoning process takes on the server. That means that some actions resulting from rule execution may be transferred back to a gateway before the reasoning process has even finished. This applies to both environments, even if the difference between the server process time and reaction time is considerably larger in the building example in contrast to the sewage scenario.

Compared to the results from the user study in section 3 we can say, that the act-mobile architecture and thus the technologies used within that architecture are well suited for small environments due to the fact that the average reaction time is 140 ms, which is not recognized as a delay by most users. In contrast the results of the building example show a more critically result. With an average reaction time of about 2500 ms and an average process time on the server of nearly 3800 ms, the execution of the reasoning process takes too long for applications like switching a light. Like shown in the user study, most users would be bothered through the resulting delay.

5   Discussion and Perspectives

While in many other publications no specific time is mentioned, our results show that the performance of a rule based reasoning task which can be extended with user defined rules is still a limitation in ontology based architectures, which underlines the findings in [12]. Hybrid approaches like proposed in [9] and [13] in turn can not be simply transferred to the act-mobile architecture because of the domain-, and thus ontology-independence. For example the separation of the static and dynamic parts of the ontology like proposed in [13] can not be adopted to the act-mobile architecture because a user may simply create an environment with a different ontology.

The problem of an insufficient performance on the reasoning task may partly be solvable through enough computation power. In addition, more optimized reasoners and triple stores could be used. However, especially in scenarios where a rule execution on a cloud server is not desirable, the trend is towards smaller devices that can be integrated into our environment. In consequence in a smart home scenario, a local gateway with low computational power may be installed to communicate with the server. Nevertheless, such a gateway should be able to execute at least some basic rules on his own to be more independent on failures of the internet connection as well as to reduce the additional latency. Accordingly, in the future work we are going to investigate the hybrid approaches in more detail to provide a way of local and fast rule execution on devices with limited resources even in a flexible architecture like act-mobile. Thus, the solution being aimed needs to be independent of a specific ontology. In addition especially for large environments, like the smart building example of our department, a distributed approach of the ontology will be considered, where for example one gateway for each floor may be installed to perform basic operations like switching a light.

6 Conclusion

The performed user evaluation showed that a delay of about 500 ms and less should be acceptable for a reaction on an event in most ambient scenarios. Based on this results the evaluation of the act-mobile architecture showed, that the architecture and thus the used technologies regarding the reasoning process are suitable for small scenarios with a limited number of triples like in the sewage treatment example. For larger environments the performance and scaleability of the reasoning task is still problematic so that further improvements are necessary for time critical applications. In consequence, in the future the development of ontology based architectures should focus even more on the performance and execution time to provide a basis for a practical use in smart environment.

References

1. Tomic, S., Fensel, A., Pellegrini, T.: Sesame demonstrator: ontologies, services and policies for energy efficiency. In: Proceedings of the 6th International Conference on Semantic Systems, I-SEMANTICS 2010, pp. 24:1–24:4. ACM, New York (2010)
2. Reinisch, C., Kofler, M., Kastner, W.: Thinkhome: A smart home as digital ecosystem. In: 2010 4th IEEE International Conference on Digital Ecosystems and Technologies (DEST), pp. 256–261 (April 2010)
3. Ausn, D., Castanedo, F., de Ipia, D.L.: On the measurement of semantic reasoners in ambient assisted living environments. In: IEEE Conf. of Intelligent Systems, pp. 82–87. IEEE (2012)
4. Horrocks, I., Patel-Schneider, P.F., Boley, H., Tabet, S., Grosof, B., Dean, M.: Swrl: A semantic web rule language combining owl and ruleml. Technical report, World Wide Web Consortium (2004)
5. Peters, M., Brink, C., Sachweh, S.: Domain independent architecture and behavior modeling for pervasive computing environments. In: International Conference on Complex, Intelligent and Software Intensive Systems, CISIS (July 2012)

6. Peters, M., Brink, C., Sachweh, S.: Including metadata into an ontology based pervasive computing architecture. In: IEEE International Conference on Ubiquitous Computing and Communications (IUCC) (July 2012)
7. Xu, J., Lee, Y.H., Tsai, W.T., Li, W., Son, Y.S., Park, J.H., Moon, K.D.: Ontology-based smart home solution and service composition. In: International Conference on Embedded Software and Systems, ICESS 2009, pp. 297–304 (May 2009)
8. Wang, X., Zhang, D., Gu, T., Pung, H.: Ontology based context modeling and reasoning using owl. In: Proceedings of the Second IEEE Annual Conference on Pervasive Computing and Communications Workshops 2004, pp. 18–22 (March 2004)
9. Agostini, A., Bettini, C., Riboni, D.: A performance evaluation of ontology-based context reasoning. In: Fifth Annual IEEE International Conference on Pervasive Computing and Communications Workshops, PerCom Workshops 2007, pp. 3–8 (March 2007)
10. Liu, C.H., Chang, K.L., Chen, J.J.Y., Hung, S.C.: Ontology-based context representation and reasoning using owl and swrl. In: 2010 Eighth Annual Communication Networks and Services Research Conference (CNSR), pp. 215–220 (May 2010)
11. Ricquebourg, V., Durand, D., Menga, D., Marine, B., Delahoche, L., Loge, C., Jolly-Desodt, A.M.: Context inferring in the smart home: An swrl approach. In: 21st International Conference on Advanced Information Networking and Applications Workshops, AINAW 2007, vol. 2, pp. 290–295 (May 2007)
12. Qin, W., Shi, Y., Suo, Y.: Ontology-based context-aware middleware for smart spaces. Tsinghua Science & Technology 12(6), 707–713 (2007)
13. Mi Park, Y., Moon, A., Il Choi, Y., Ki Kim, S., Kim, S.: An efficient context model for fast responsiveness of context-aware services in mobile networks. In: 2010 7th IEEE Consumer Communications and Networking Conference (CCNC), pp. 1–5 (January 2010)
14. Urbani, J., Kotoulas, S., Maassen, J., van Harmelen, F., Bal, H.: OWL reasoning with webPIE: Calculating the closure of 100 billion triples. In: Aroyo, L., Antoniou, G., Hyvönen, E., ten Teije, A., Stuckenschmidt, H., Cabral, L., Tudorache, T. (eds.) ESWC 2010, Part I. LNCS, vol. 6088, pp. 213–227. Springer, Heidelberg (2010)
15. Ren, Y., Pan, J.Z., Lee, K.: Parallel aBox reasoning of $\mathcal{EL}$ ontologies. In: Pan, J.Z., Chen, H., Kim, H.-G., Li, J., Wu, Z., Horrocks, I., Mizoguchi, R., Wu, Z. (eds.) JIST 2011. LNCS, vol. 7185, pp. 17–32. Springer, Heidelberg (2012)
16. Stavropoulos, T.G., Vrakas, D., Vlachava, D., Bassiliades, N.: Bonsai: a smart building ontology for ambient intelligence. In: Proceedings of the 2nd International Conference on Web Intelligence, Mining and Semantics, WIMS 2012, pp. 30:1–30:12 (2012)

Using 3D Virtual Agents to Improve
the Autonomy and Quality of Life of Elderly People

Piedad Garrido[1], Angel Sanchez[1], Francisco J. Martinez[1], Sandra Baldassarri[2],
Eva Cerezo[2], and Francisco J. Seron[2]

[1] EUPT, Ciudad Escolar s/n
{piedad,asespilez,f.martinez}@unizar.es
[2] EINA, María de Luna, 5
{sandra,ecerezo,seron}@unizar.es

Abstract. Nowadays, the percentage of elderly people is increasing, especially in developed countries. Technological products can be used to propel a cohesive and inclusive inter-generational society, although they should be adapted to satisfy the needs and preferences of elderly people. In this paper, we present 3D virtual agents as a promising system to improve the autonomy and Quality of Life (QoL) of elderly people, as they are technological systems that look and act like a person, thanks to the use of speech recognition and voice communication, and they can facilitate users' everyday life. In this work, the connection between a virtual agent with a SQL database engine allows our agent to be connected to the health center, thus being able to assist users in their daily intake of medication.

1 Introduction

In last decades, there has been a considerable increase in the average age of the inhabitants and the life expectancy, especially in developed countries. According to a statistical study done by Eurostat [5], the percentage of population in the European Union (EU) aged 65 years or older was of 22.5% in 2005, and it is expected to increase to 30% in 2050. Nowadays, in Spain, 17.4% of the total population is over 65 years old, according to data collected by the Spanish National Statistics Institute; additionally, they estimate that this percentage will reach 32% in 2049.

Statistical data are even more revealing if we focus on rural areas, which present special demographic characteristics, such as: (i) the gradual depopulation of its territory, (ii) aging population, (iii) declining birth rates, and (iv) increasing life expectancy.

The purpose of this work, based on the use of Information and Communications Technologies (ICTs) in rural areas, is to create a system that tries to improve the autonomy and quality of life for the elderly people. Our objective is to implement a system able to assist elderly people with their day-to-day activities, making their life more comfortable. In particular, our approach focuses on covering the needs that have been identified in this collective related to health and safety, specifically, to assist elderly people in their daily tasks (e.g., their intake of medication). These needs are more important in depopulated areas with aging population [11], since these areas are prone to present a lack of resources and opportunities for their inhabitants, compromising people's autonomy, and thereby provoking a loss of quality of life.

A. van Berlo et al. (Eds.): *Ambient Intelligence – Software & Applications*, AISC 219, pp. 129–136.
DOI: 10.1007/978-3-319-00566-9_17   © Springer International Publishing Switzerland 2013

Table 1. Population Evolution in the last years (source: http://www.ine.es)

	1999	2002	2005	2008	2011
Maestrazgo	3,717	3,700	3,739	3,789	3,670
Teruel	136,849	137,342	141,091	146,324	144,607
Aragon	1,183,234	1,217,514	1,269,027	1,326,918	1,346,293
Spain	40,202,160	41,837,894	44,108,530	46,157,822	47,190,493

Based on a study which reviewed some important issues regarding the quality of life of older people in rural areas [9], we focus on the Maestrazgo, a rural area of the province of Teruel (Aragon, Spain), whose demographic statistics reveal that, while the rest of Spain is gaining population, this region continues with the same figures during the last 13 years (see Table 1). Specifically, 28.6% of the Maestrazgo population is over 65 years, in contrast to Aragon, which has a 20%, or Spain, which has a 17.4%.

In this work we use Maxine, a virtual agent platform [1], developed by the Advanced Computer Graphics Group (GIGA) of the University of Zaragoza. In particular, we have implemented a prototype able to connect online 3D virtual agents with the health centers of the province of Teruel, in order to carry out more humanized telemedicine and telecare tasks (e.g., the correct daily intake of medication, and task reminder), functionalities that currently are performed by using cellphones or tracking devices. The software responsible for making the connection is integrated into the agent's knowledge base and it makes use of modern techniques of Artificial Intelligence (AI) based on natural language processing and automatic speech recognition, since we want to provide users an easy way to interact with our system in a more natural way by means of a user-friendly interface.

The rest of the paper is structured as follows. The following section explains the background and the different needs that motivated our work. Section 3 describes the Maxine's architecture and how user interacts with Maxine. Later, Section 4 describes our prototype, detailing: (i) the procedure of via voice communication with the agent, (ii) the query process, and (iii) the intake of medication alarms. Finally, Section 5 presents the conclusions and future work.

2 Background and Motivation

Quality of life (QoL) is a concept that evaluates the general welfare of individuals and societies. The term is used in many contexts and fields, such as sociology, political science, medicine, development studies, etc. Our proposal is based on the analysis and measurements, centered in QoL for elderly people, made by authors such as Brown et al. [3], and Bowling and Stenner [2], as well as on the needs identified in the study focused on the autonomy and QoL of elderly people in the Maestrazgo [9]. After identifying the factors that these works have in common, we developed our own QoL model, which has been called 'Common Working Model' (CWM), which aims to reflect the requirements that allow elderly people to remain in their homes safely, according to their preferences and desires.

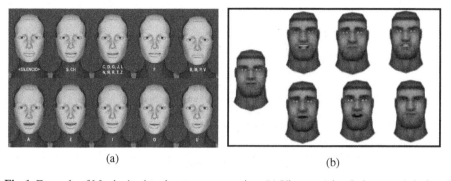

(a)   (b)

Fig. 1. Example of Maxine's virtual actors representing: (a) Visemes (visual phonemes) designed for the lip-sync in Spanish, and (b) basic universal facial expressions.

Our CWM accounts for the following factors: health, psychological and physical welfare, and social integration and adaption. The prototype developed covers all of these factors, as follows: (i) as for health, it reminds users their daily intake of medication, and manages their weekly pill box, (ii) regarding psychological welfare, users interact with a "humanoid" interface through voice, (iii) as for physical welfare, the user does not need to memorize anything, and does not require previous training, and (iv) regarding social integration and adaption, as it assists people in their daily activities, they can remain in their homes more time.

Although the market offers other initiatives based on the use of ICTs to facilitate access to health services through Internet, Digital Terrestrial Television (DTT), or cellphones, the use of Embodied Conversational Agents (ECAs) in this area is not widespread, and it is often centered on domotics [6], or aimed at people who suffers from some type of chronic disease such as dementia, or Alzheimer's disease [10]. In this paper we show the use of an interactive 3D animated agent as a new user-friendly interface in geriatric environments. The main advantage of our work compared with others, is that the user interacts with a "humanoid" interface via voice. The 3D virtual agent is provided with both expressivity (facial) and emotional capabilities. Our system interacts with the user naturally, as he/she can directly talk to the virtual agent through natural language questions without requiring to memorize anything, or to be previously trained.

3   Maxine: The Animation Engine

3.1   Overview

Maxine is a powerful engine for the management, animation and visualization of 3D worlds [1]. It allows not only the creation and management of objects, scenes and virtual characters, but also real-time multimodal and emotional interaction with users. The scenarios can include 3D virtual actors endowed with facial and body emotion expression, as well as lip-sync animation and emotional synthetic voice (see Figure 1). Virtual actors can also vary their answers during a conversation depending on the relationship with the user, so that they are especially suitable for their use as natural interfaces.

```
<aiml version="1.0.1">
    <category>
        <pattern>MEDICINES</pattern>
            <template>
                <random>
                <li>Give me your ID, please</li>
                <li>Please, your ID?</li>
                </random>
            </template>
    </category>
    <category>
        <pattern>My ID is:</pattern>
            <template>
            <think>
                <set name="ID">
                <star/></set>
                <script>:QUERY</script>
            </think>
            </template>
    </category>
```

Fig. 2. Example of AIML code

The engine has been written in C++ and employs a set of open source libraries, excepting the libraries required for the speech recognition and generation capabilities that are carried out through the Loquendo[1] commercial software.

3.2   User Interaction with Maxine

In order to allow conversations in natural language with the virtual characters, Maxine is able to recognize user's speech. For this purpose, our platform includes the Automatic Speech Recognition (ASR) system of Loquendo, which allows to process any question, thus generating the appropriated answers in real time, due to the support and fulfilment of the following standards: W3C SRGS (Speech Recognition Grammar Specification), ABNF (Augmented Backus-Naur Form), and SISR (Semantic Interpretation for Speech Recognition). The answers of the system are communicated to the users through the Loquendo TTS (Text-to-Speech) voice synthesizer, which satisfies the W3C SSML (Speech Synthesis Markup Language), and therefore allows to obtain a simple and natural dialogue that simulated the human voice in a very natural way.

Furthermore, Maxine allows to create and manage dynamic knowledge databases with their corresponding questions and answers. The performance of the answering engine and the syntax used for the definition of a database is based on the Artificial Intelligent Markup Language (AIML), a markup language that derives from XML, with the aim of being served, received and processed in the web. In this work, the AIML code used in Maxine has been modified to allow the execution of embedded SQL scripts required to obtain the medication needed, or to manage the alarms (see Figure 2).

---

[1] http://www.loquendo.com

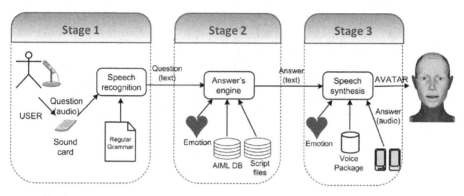

Fig. 3. Stages of the voice communication process between the user and the agent

4 Description of Our Prototype

Based on the deficiencies observed in the QoL and the problems showed by elderly people that lives in rural areas, we consider to apply the use of Information and Communications Technologies in order to overcome these problems.

The prototype developed is focused, on the one hand, in the use of Embodied Conversational Agents (ECAs), which are software entities that have human appearance, that act like human beings, and that are able to establish a conversation, or collaborate when carrying out different tasks [4]. On the other hand, the prototype is characterized by the development of a innovative software that allows connecting the Maxine's answering engine, designed in AIML, with a relational answering engine.

The innovation in this project is centered in the connection between an ECA, an entity from a virtual world [7], with an engine for searching answers, based on the use of the SQL standard language. This relational database is in charge of remembering the correct intake of medication through the alarms that are communicated via voice to the users, informing about the hour, type, and doses of each medication.

Our system allows that the health center can create the alarms adapted to each patient, and that the user can ask the virtual agent about the medication he/she has to take each day. In order to obtain the required information, the virtual agent accesses to the relational database of the health center, obtains the list of the daily alarms, and communicates the response to the user, interacting with him in natural language.

In order to facilitate users the management of the system, neither they nor their relatives have to introduce the information related with the medication. The health center is directly in charge of providing this information, through a simple online application for introducing the data into the database. Besides helping users to remind their daily medication, our system interacts with the user, since he/she can talk or ask questions to the virtual agent. Therefore, the system developed improves the QoL of elderly people, satisfying all the factors of our Common Working Model (CWM).

4.1 Communication between the User and the Virtual Agent

The communication process between the user and the virtual character is performed via voice, and it is divided into three stages [1], as shown in Figure 3:

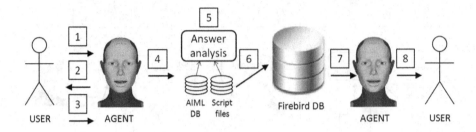

Fig. 4. Process of medication request

1. The user formulates an order or question in natural language. The audio generated is picked up by the microphone and the sound card. The ASR software, using a grammar with the dictionary of all the words that can be recognized, converts the audio input into a text with the words pronounced by the user.
2. The search engine processes the meaning of the input, assigning an answer to that question, either looking for it in the AIML database, or generating it from the execution of a script.
3. The speech synthesis is performed with the Spanish voice packages offered by Loquendo TTS. Additionally, SAPI [8] is used for gathering information regarding the visemes corresponding to each word. This information is necessary for performing the virtual character lip synchronization. In this way, the user can hear the answers, and at the same time, see the humanized virtual character.

4.2 Application: Assistance with the Medication

Figure 4 describes the communication process between the user and the virtual agent in order to obtain the information required. In particular, it shows the process from the moment that the user asks a question, until the virtual agent indicates the medication to take. The steps followed in the process are:

- Step 1: The user, for example, formulates the following question to the virtual character: "Which medication I have to take?"
- Step 2: The virtual agent analyzes the question and answers: "Give me the number of your ID card"
- Step 3: The user communicates the required data to the virtual agent.
- Step 4: The virtual agent receives the information, performs the speech recognition process, and sends it to the module in charge of analyzing questions and answers.
- Step 5: This module decides the answer that must give to the user. This information is not stored in the knowledge database of the agent, but in an external relational database implemented in FireBird, hence, a communication channel between the two knowledge databases must be created.
- Step 6: Using the communication channel created by a Java servlet, the virtual agent sends the information about the ID card to the database, together with the present date, and it makes a SQL request about all the alarms that the user has for that date.

- Step 7: Using the same communication channel, the alarms obtained in SQL are automatically sent, filtered, and parsed to XML, thus making the result AIML compatible.
- Step 8: The virtual agent automatically processes the information, and through its voice, communicates the different alarms to the user. If there are no alarms, the agent informs user that he/she does not have to take any medication.

5   Conclusions and Future Work

In this paper we have presented a 3D virtual agent able to improve the QoL of elderly people. Its main objective is to remind users when and how take their medication. Additionally, it can mitigate their excessive loneliness, one issue quite common in aging societies. Our intention has been to create a system closer to the user; the interaction between the user and the virtual agent is made in natural language. In the future, we consider to enhance our system in several ways: (i) using the open source Julius speech recognition software instead of Loquendo, (ii) using Web services instead of servlets for the communication between the virtual agent and the relational knowledge base, thereby improving system performance (providing higher speed and independence, and a more efficient information retrieval process), and (iii) adapting the interface for the implementation in other devices.

Acknowledgement. The authors would like to acknowledge the Fundación Universitaria Antonio Gargallo (FUAG), and the Caja de Ahorros de la Inmaculada (CAI), for their support to this project. This work has also been partially supported by the Spanish "Dirección General de Investigación", under the contract number TIN2011-24660.

References

1. Baldassarri, S., Cerezo, E.: Maxine: Embodied conversational agents for multimodal affective communication. In: Mukai, N. (ed.) Computer Graphics, ch. 11. In Tech Open-Access Publishing (2012)
2. Bowling, A., Stenner, P.: Which measure of quality of life performs best in older age? A comparison of the OPQOL, CASP-19 and WHOQOL-OLD. Journal of Epidemiology & Community Health 65, 273–280 (2011)
3. Brown, J., Bowling, A., Flynn, T.: Models of Quality of Life: A Taxonomy, Overview and Systematic Review of the Literature. Technical report (2004)
4. Cassell, J., Sullivan, J., Prevost, S., Churchill, E.F.: Embodied conversational agents. MIT Press, Cambridge (2000)
5. Eurostat. The life of women and men in Europe - a statistical portrait (2008)
6. Fiol-Roig, G.: Intelligent agent for home assistance. In: Proceedings of The Eleventh IASTED International Conference on Artificial Intelligence and Soft Computing (ASC 2007), Anaheim, CA, USA, pp. 102–107 (2007)
7. Garrido, P., Martinez, F.J., Guetl, C., Plaza, I.: Enhancing intelligent pedagogical agents in virtual worlds. In: Workshop on Virtual Worlds for Academic, Organizational, and Life-long Learning (ViWo 2010), held with ICCE, Putrajaya, Malaysia (November 2010)
8. Long, B.: Speech synthesis and speech recognition using SAPI 5.1. In: Delphi Developers Conference (DCon 2002), Reading, England (June 2002)

9. Rubio, M., Plaza, I., Hernanz, C.: Autonomía y Calidad de Vida de los Mayores en el Maestrazgo. D.L. TE-21/2011. Aragón Vivo, Artes Gráficas (2011)
10. Sakai, Y., Nonaka, Y., Yasuda, K., Nakano, Y.I.: Listener agent for elderly people with dementia. In: Proceedings of the seventh Annual ACM/IEEE International Conference on Human-Robot Interaction (HRI 2012), pp. 199–200. ACM, Boston (2012)
11. Sasaki, J., Yamada, K., Tanaka, M., Funyu, Y.: An experiment of life-support network for elderly people living in a rural area. International Journal of Systems Applications Engineering & Development 1(4), 120–124 (2010)

Multimodal Indoor Tracking
of a Single Elder
in an AAL Environment*

Andrei-Adnan Ismail and Adina-Magda Florea

University Politehnica of Bucharest, 060042, Bucharest, Romania
andrei.ismail@cs.pub.ro
http://aimas.cs.pub.ro/people/andrei.ismail

Abstract. In this paper we present a proof-of-concept practical architecture of a system for tracking older adults living alone in an ambient intelligence environment by using multi-modal sensory information, such as audio and video input from Kinect and Arduino devices. This AAL (Ambient Assisted Living) system collects data into a Database of Trajectories, used to automatically generate training examples. We then show how the Database of Trajectories can be used in conjunction with a set of well-trained algorithms to alert remote caregivers in case of an unfortunate event, such as the elder falling down - one of the leading causes of injury and death.

1 Introduction

The world's population is ageing: studies project that, by 2050, over 32% of people in industrialized countries and 20% in developing countries will be over 60 [5]. As part of the public policy-making mechanism, increased attention was given to developing intelligent environments that can safely accomodate elder with physical and/or mental problems and alert caregivers in case of important events. The United Nations have published "United Nations Principles for Older Persons", in which one of the key elements is safe environments, enabling the elderly to live at home independently for as long as possible.

Monitoring and real-time data processing have a key role in building a smart environment that is able to prolong the period of independence of an elder living alone at home. While working on the problem of tracking an elder at home, we have identified the following challenges:

C1 : the environment may be cluttered with many different objects accumulated throughout life (this depends very much on the culture and the country; people in developing countries tend to have more cluttered apartments).

---

* This work was supported by projects ERRIC - Empowering Romanian Research on Intelligent Information Technologies/FP7-REGPOT-2010-1, ID: 264207 and "DocInvest" - POSDRU/107/1.5/S/76813.

A. van Berlo et al. (Eds.): *Ambient Intelligence – Software & Applications*, AISC 219, pp. 137–145.
DOI: 10.1007/978-3-319-00566-9_18   © Springer International Publishing Switzerland 2013

C2 : the tracking system must be non-invasive and easy to train (the monitored person should not be required to wear anything or to press any buttons). Also, for training, it should have a simple way to accomodate temporary visits of a relatively large number of persons without setting off an alarm.

C3 : this system has to work 24/7 in actual living conditions. In particular, it cannot expect to have proper lighting during the night. Therefore, it should use infrared capabilities and microphones in order to compensate in these cases.

C4 : adding a new sensor to the system should be simple and the existing sensors should be main-stream, so that the it is easy to scale depending on the available funds. Also, data processing should be done in the cloud, where cheaper and more reliable computing resources are available.

As the MIT House_n consortium team note in their work on challenges of deploying smart systems [12], there is a conflict of interests between the user perception of certain types of sensors (video cameras and microphones) and the amount of information needed by the system in order to do meaningful inference. On the one hand, users prefer "non-invasive" means of detecting activity such as RFID tags and proximity sensors, but these are needed in larger numbers, harder to set-up, train and more expensive to operate in the long run. [12] also mentions 12 questions that prospective research work on intelligent homecare systems should answer related to practical aspects of their deployment, training and maintenance, aspects largely ignored by the currently published work.

We propose a proof-of-concept in-door tracking system developed in the spirit of these questions, that monitors a single elder at home. This tracking system will provide permanent information on the approximate physical location of the tracked person and its body posture by combining audio and video information streams from off-the-shelf mainstream components. Published work on combining audio and video in tracking systems [7] confirms our assumption that it improves accuracy. Even though this system does not tackle the problems of activity or sentiment detection, it can serve as a solid building block for such an undertaking.

The outline of this paper is the following: in section 2 we will present the general system architecture. In section 3, we will present the chosen hardware components and the actual physical layout of the proposed system, providing estimated costs and a list of unexpected challenges. Section 4 describes our processing infrastructure and a model for storing information related to tracking a person in space and time. We demonstrate how this data structure can be further used to detect whether the person has fallen down. In section 5, we present our conclusions and future research directions.

2   Related Work

The Kinect is a revolutionary sensor given its sensing quality, capabilities and price tag. It has spanned a large number of indoor-related projects such as

3D modelling of indoor environments [4] and multiple amateur-level enthusi-ast projects posted on websites such as http://www.kinecthacks.com. Also, multiple Kinects have been shown to successfully work together despite some physical interference among them in [3].

Previous work on tracking the location of a person in an indoor setting can roughly be divided into two large categories: related to estimating future loca-tion of the subject by using a statistical model of the activity, while the opposite approach is using as much information as possible with sensors emulating the human senses embedded into the environment. Singla et al. use in [11] a combi-nation of proximity sensors and sensors placed on objects such as taps, together with Markov models, in order to estimate the current activity, and thus, the most probable next location(s), while PlaceLab [8] uses over 300 sensors, featuring full video, audio and infrared coverage of an apartment-scale laboratory, being closer to our own approach but with a much more intrusive set-up (we have optimized ours so that it can be installed without too much trouble in any reasonably-sized home: it just needs some extra electrical wires and network connections).

Our work complements these approaches by taking what is best in each: the use of multiple Kinects, in a simple and cost-efficient set-up that does not make too many assumptions about the operating environment (for example, night tracking is rarely mentioned in indoor tracking literature).

A similar system is currently being developed in Austria - "CARE: Safe Pri-vate Homes for Elderly Persons" [1], but it does not feature multi-modal infor-mation processing and correlation.

3   General Architecture of the Tracking System

The general architecture of our proposed tracking system is presented in figure 1 (page 140).

The sensing infrastructure is built on top of a number of Microsoft Kinect sensors and Arduino boards, from which data is fetched using either a wired USB connection (for high-bandwidth measurements such as frames), or a wireless 802.11/b connection (for discrete measurements such as the output of an infrared proximity sensor).

The software that processes this data is organised as a pipeline made up of individual PDUs (processing data units) communicating through a system of distributed queues. There are two types of PDUs: data acquisition PDUs, which poll the sensors for new data and record these measurements into a Database of Trajectories, and data crunching PDUs which use face recognition and voice recognition webservices and the recorded data to generate automatic examples by correlating spatio-temporal information.

The software is installed on 2 types of processing nodes: data acquisition nodes (running data acquisition PDUs and the database engine), and data crunching PDUs (running detection and correlation algorithms).

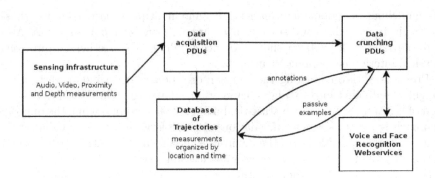

Fig. 1. Diagram of the tracking system

4   Hardware Components and Physical Layout

Our tracking system uses only off-the-shelf sensing components:

- **9 Microsoft Kinect** sensors. Priced as low as $100, these contain an RGB camera and a depth sensing device and are able to deliver at 30 frames per second (fps) both RGB and depth information. The SDKs available both on Linux and on Windows contain well-trained algorithms able to extract the positions of 17 joints, including head, neck and shoulders.
- **18 Arduino Mega boards**, with the following configuration:
  - Atmel AtMega 2560 8-bit processor running at 16 MHz
  - WiFly shield that enables wireless communication (802.11 b/g)
  - Sharp infrared proximity sensor (sensing range: 20-150 cm)
  - Electret Microphone

  One such completely equipped board is priced around $250.

The tracking system gathers the following type of data from the sensors:

Sensing Equipment	Type of Information Gathered
Microsoft Kinect	– Joint Position (2D/3D coordinates of the tracked person's skeleton joints) – RGB image (1280 x 1024 x 32bpp) – Depth image (640 x 480 x 8 bpp)
Arduino with WiFly shield	– Estimated distance to nearby person – audio recording (useful to infer sound source position)

The total of 27 equipments is grouped into 9 T-shaped *keypoints* numbered $K1, K2, \ldots, K9$, each of them containing one Kinect and 2 Arduino equipments. The actual physical layout of a keypoint is given in figure 2 (page 141).

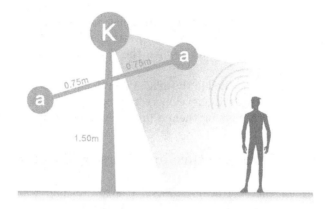

Fig. 2. Keypoint - a group of 1 Kinect (k) and 2 Arduino (a) sensing equipments

The 9 keypoints $(K1 - K9)$ were placed in our 8.5 m × 4.5 m room as evenly spaced as possible, given the constraints (door, windows and cabling). Their approximate layout is specified in figure 3 (page 141).

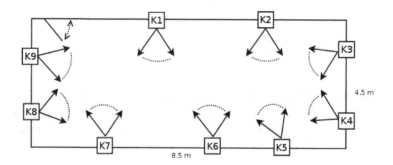

Fig. 3. Approximate layout of the 9 keypoints in the experiment room

The total cost of the prototype is under $10k$, a promising answer to question 3 from [12]: "What is the cost of end-user installation?". We prefer cost-effective information-complete sensors to privacy-sensible deep sensing approaches.

Data processing is done using local servers, planning to move to the cloud in the future, thus trading a known and measurable extra latency for availability.

As a final note on the hardware part of the system, three unexpected important details must be mentioned:

- the data acquisition node must be near the keypoint. The USB 2.0 standard [2] notes a maximum length of 3.5 m for an USB cable, and the Kinect video data and Arduino audio data can only be fetched via the USB cable (due to Arduino WiFi driver limitations).
- Arduino proximity information is fetched using a WiFi connection and is supposed to complement the RGBD information from the Kinects. While the extended range for the Kinects is pretty generous (0.7 m − 6 m), it is clear that they cannot be used when the tracked person is too near the sensors
- even though it is not specified in the official user documentation, independent tests from multiple hardware developers have confirmed that Kinect sensors also work in the dark, making them a proper way to tackle challenge $C3$.

5   Processing Infrastructure and Person Tracking

Data processing for such a tracking system has several important properties. First of all, data loss for short periods of time is not an issue and we can even contemplate sampling it in order to reduce costs. Secondly, a latency in the order of seconds is acceptable in delivering the result of processing. If a known person enters the room and the system greets it 5 seconds later, the perceived effect is more than acceptable (however, this delay must be always bounded).

Processing the stream of data using PDUs is an industry-standard approach and has several advantages:

- high-availability by having hot replacements for PDUs in place, managed by a supervisor process
- easy vertical scaling by distributing different PDUs on different machines (a partial solution to challenge C4)
- the ability of throwing away messages from the queues as the cost of maintaining a bounded delay in emergency situations

Examples of PDUs include: DatabaseWriterPDU (writes all measurements to a database), HeadCropPDU (crops headshots from a stream of incoming raw frames), FaceRecognitionPDU (identifies a given face against a list of known person).

5.1   The Database of Trajectories

All PDUs have access to a shared database of person trajectories. This database Δ should implement the following operations:

- **start a new tracking session,** Δ.startNewSession(person_identity).
  Whether the person is known to the system or not, a new session will be started when a person is detected. Unknown persons will be assigned a random ID and the caregiver will be given the option to identify them at some point.

- **add a location to a trajectory,**
  Δ.addLocation(person_identity, (X, Y, Z))
  This means that a PDU has detected that the specific person is at the position (X, Y, Z).
- **retrieve the trajectory of a given person,** Δ.trajectory(person_identity).
- **retrieve the persons nearby a given position,** Δ.personsNear(X, Y, Z).
- **annotate a location from a trajectory with additional information,**
  Δ.annotateLocation(person_identity, (X, Y, Z), info)
  After detecting that a person is at the location (X, Y, Z), one or more PDU's are able to bring actual proof that the person is at that location.
  What kind of information can serve as an annotation to a location of a trajectory? It can be a sample voice taken from a nearby microphone, the joint coordinates of the skeleton of the person, a headshot of the nearby person and many more.

The purpose of this database is to build a list of successive annotated positions for a given person, where annotations serve as proofs that the person was actually there, and can be correlated to solve challenge C2.

The Database of Trajectories can be used to generate training examples automatically: if person P enters the room but is not identified for the first 300 seconds due to a new haircut and sunglasses, when a PDU finally identifies the person using its voice, it can mark previous headshots of the person as belonging to it. We believe that self-generating passive examples are key to keeping a tracking system up-to-date. This scenario is described formally in algorithm 1 (page 144).

5.2 Detecting When the Person Has Fallen Down

In the last section we have defined the database of trajectories, and shown how it works like a time and location based index of person knowledge.

Our end goal is alerting the caregiver of the elder when a fall has occured. An exhaustive classification of other unfortunate events and action plans can be found in a novel, freely editable format in [9].

Given our network of sensing devices, we propose the following novel conditions for detecting a fall:

- if the person was being tracked by one of the Kinects, the center of gravity of the person's skeleton has a large downward acceleration and afterwards the person cannot be tracked anymore, followed by a falling noise or by a scream. The center of gravity is computed as the average over all 17 joint 3D coordinates provided by the Kinect.
- if the person was not being tracked by one of the Kinects, but was in the visual range of one, the person's contour can be computed using background subtraction [10], and then the center of gravity can be approximated from this contour. This condition consists of a large downward acceleration followed by a falling noise or by a scream.

Data: a voice sample s from microphone $M1$
Result: a set $(F_1, F_2, \ldots, F_n)$ of headshots of the same person
fed as examples to the face recognition algorithm

```
person_identity = voiceRecognition(s);
if person_identity ≠ NULL then
    nearby_persons = Δ.nearbyPersons(X_M1, Y_M1, Z_M1);
    if len(nearby_persons) > 1 then
        # cannot infer anything, so give up;
        return [];
    end
    # Retrieve the frames from the trajectory containing the person;
    frames = [t.info.frame for t in Δ.trajectory(person_identity)];
    # Extract headshot by detecting face and possibly correlating with skeleton;
    headshots = [cropHeadshot(frame) for frame in frames];
    # Feed the headshots to the face detection algorithm as examples;
    forall the headshots do
        addFaceRecognitionExample(headshot, person_identity);
    end
end
```

Algorithm 1. Learning Face Samples Starting From a Voice Sample

Our proposal can even solve challenge C1 to some extent: even if the room is cluttered with objects, the person will not be completely surrounded by such objects. So he/she will be in the range of one Kinect camera (possibly facing backwards), increasing the probability of triggering our conditions.

Even if the light is turned off, the Kinect can still track skeletons in the dark, and this can be complemented by background subtraction on the depth image.

6   Conclusions and Future Work

We have presented a practical architecture for tracking an elder in an indoor environment and detecting falls in a holistic way: from the necessary hardware, to estimated costs, to an outline of a software architecture and to a novel way of organizing information recorded by it. Such a tracking system can reduce the damaging effects of falling (one of the leading causes of death and injury in elder [6]) by enabling a quick reaction.

The system depicted in this work has been partially implemented in a laboratory of Politehnica University of Bucharest. Primary experiments are showing a propagation time of at most 20 seconds from the moment of data acquisition to the actual recognition of the user, even when using 3rd party APIs that run in the cloud. Also, we have performed reproducible tests for our processing pipeline by performing a sensor data dump during a period of time and running experiments on the recorded data.

This contribution lays the foundation for building intelligent systems that react to unfortunate events within a bounded timeframe. The Database of Trajectories model can be extended to support annotations in RDF format that store results of algorithms such as object detection. This would also allow integration between the database and public triple stores such as DBPedia, containing common sense knowledge. One important extension of this system would be a playback infrastructure that delivers measurements without having an actual physical infrastructure - this would allow research in AAL without having actual access to a well-equiped laboratory.

References

1. Catalogue of projects 2012. Ambient Assisted Living Joint Programme
2. Universal Serial Bus Specification Revision 2.0 (2011) (retrieved September 8, 2012)
3. Berger, K., Ruhl, K., Brümmer, C., Schröder, Y., Scholz, A., Magnor, M.: Markerless motion capture using multiple color-depth sensors. In: Proc. Vision, Modeling and Visualization (VMV), vol. 2011, p. 3 (2011)
4. Henry, P., Krainin, M., Herbst, E., Ren, X., Fox, D.: Rgb-d mapping: Using kinect-style depth cameras for dense 3d modeling of indoor environments. The International Journal of Robotics Research 31(5), 647–663 (2012)
5. Bloom, D.E., Fried, L.P., Hogan, P., Kalache, A., Beard, J.R., Biggs, S., Jay Olshansky, S. (eds.): Global Population Ageing: Peril or Promise. World Economic Forum, Geneva (2011)
6. Kannus, P., Parkkari, J., Niemi, S., Palvanen, M.: Fall-induced deaths among elderly people. Journal Information 95(3) (2005)
7. Kushwaha, M., Oh, S., Amundson, I., Koutsoukos, X., Ledeczi, A.: Target tracking in heterogeneous sensor networks using audio and video sensor fusion. In: IEEE International Conference on Multisensor Fusion and Integration for Intelligent Systems, MFI 2008, pp. 14–19. IEEE (2008)
8. Logan, B., Healey, J.: Sensors to detect the activities of daily living. In: 28th Annual International Conference of the IEEE Engineering in Medicine and Biology Society, EMBS 2006, pp. 5362–5365. IEEE (2006)
9. Lyons, P., Cong, A.T., Steinhauer, H.J., Marsland, S., Dietrich, J., Guesgen, H.W.: Exploring the responsibilities of single-inhabitant smart homes with use cases. Journal of Ambient Intelligence and Smart Environments 2(3), 211–232 (2010)
10. Piccardi, M.: Background subtraction techniques: a review. In: 2004 IEEE International Conference on Systems, Man and Cybernetics, vol. 4, pp. 3099–3104. IEEE (2004)
11. Singla, G., Cook, D., Schmitter, M.: Edgecombe. Incorporating temporal reasoning into activity recognition for smart home residents. In: Proceedings of the AAAI Workshop on Spatial and Temporal Reasoning, pp. 53–61 (2008)
12. Kaushik, P., Intille, S.S., Rockinson, R.: Deploying context-aware health technology at home: Human-centric challenges. In: Human-Centric Interfaces for Ambient Intelligence, pp. 479–503. Elsevier (2010)

Distributed Neural Computation over WSN in Ambient Intelligence[*]

Davide Bacciu[1], Claudio Gallicchio[1], Alessandro Lenzi[1], Stefano Chessa[1,2],
Alessio Micheli[1], Susanna Pelagatti[1], and Claudio Vairo[2]

[1] Dipartimento di Informatica, University of Pisa, largo Pontecorvo 3, 56127 Pisa, Italy
{bacciu,gallicch,lenzi,ste,micheli,susanna}@di.unipi.it
[2] ISTI-CNR, via Moruzzi 1, 56124 Pisa, Italy
vairo@isti.cnr.it

Abstract. Ambient Intelligence (AmI) applications need information about the surrounding environment. This can be collected by means of Wireless Sensor Networks (WSN) that also analyze and build forecasts for applications. The RUBICON Learning Layer implements a distributed neural computation over WSN. In this system, measurements taken by sensors are combined by using neural computation to provide future forecasts based on previous measurements and on the past knowledge of the environment.

Keywords: Ambient Intelligence, Wireless Sensor Networks, Neural Networks.

1   Introduction

Wireless Sensor Networks (WSN) [1] play an important role in Ambient Intelligence (AmI) [2], since it is one of the primary sources of real-time, contextual information to be used in decision-taking processes. However, their role is typically limited to data acquisition with minimal data filtering and preprocessing. This approach to the use of WSN in AmI has been motivated by the limited computational resources of the sensors and by the complexity of sensors programming, which requires a deep knowledge of embedded systems and wireless communications [1]. On the other hand, sensors have data filtering and preprocessing capabilities that, although limited, are still unexplored. These features are of extreme importance to AmI applications, since smarter sensors may become a permanent part of the environment where they are deployed and learn to recognize events typical of that environment. In this way, a WSN may provide information on the environment to mobile components (such as robots or smart phones) that are not permanently included in the system: based on its past experience it can provide a description of the environment and of the occurring events. In other words, the WSN is part of an ecology, along with robotic companions, robots, smart phones, appliances etc., where all elements of the ecology implement a distributed learning system. This is the focus of the EU FP7 project

---

[*] This work is supported in part by the EU FP7 RUBICON Project (contract no. 269914).

A. van Berlo et al. (Eds.): *Ambient Intelligence – Software & Applications*, AISC 219, pp. 147–154.
DOI: 10.1007/978-3-319-00566-9_19   © Springer International Publishing Switzerland 2013

RUBICON [3], and specifically of the RUBICON Learning Layer (LL), whose architecture and implementation is presented in this work.

2   Related Works

Learning in WSNs introduces unseen challenges with respect to adaptive control and data processing, mostly due to computational limitations and distributed architecture of WSN. Artificial Neural Networks (ANNs) [4] are one of the most important learning paradigms, characterized by interesting analogies with the distributed nature of WSNs. Recent applications of ANN to WSN can be found in [5], [6], [7]. However, current work on neural applications to WSNs is fairly limited in exploitation of the distributed sensor architecture. Learning techniques are used to find approximated solutions to very specific tasks, mostly within static WSN configurations. Learning solutions have a narrow scope, resulting in poor scalability, given that different tasks are addressed by different learning models. Further, these solutions are centralized or characterized by little cooperation between the distributed learning units [8]. Recently, Reservoir Computing (RC) [9], [10] has gained increasing interest as modeling paradigm for Recurrent ANN (RNN). Within this framework, we take into account the Echo State Network (ESN) model [11]. A typical RC network is composed of a dynamical component called *reservoir*, and a static component called *readout*. The former implements a recurrent contractive encoding of the driving input sequences into a fixed size state space, providing the readout with a large "reservoir" of dynamics to linearly combine for output computation. The RC/ESN approach is rather efficient, as the readout is the only trained part of the network, while the reservoir is initialized under contractive constraints [12] and then is left untrained. Overall, the RC paradigm is able to conjugate the power of RNN in capturing dynamic knowledge from sequential information with the computational feasibility of learning by linear models. Thereby, RC networks result in suitable models for learning in the computationally constrained WSN scenario and for being embedded on-board the sensor nodes. RC application to WSN for user's indoor movements forecasting has been proposed in [13], [14], showing the appropriateness of the RC models when dealing with the complex dynamics of data gathered by WSNs.

As compared to previous works, in the design of the RUBICON Learning Layer (LL) we take a different, innovative approach in the exploitation of RC in a distributed scenario comprising a loosely coupled network of computationally constrained devices. Specifically, we realize a general purpose learning system capable of addressing a large variety of computational learning tasks through a scalable distributed architecture comprising independent RC learning modules deployed on a variety of devices, including sensors. The LL provides learning services through a distributed neural computation that, differently from the works in literature, is capable of catering for the dynamicity of both the monitored environment as well as of the underlying network.

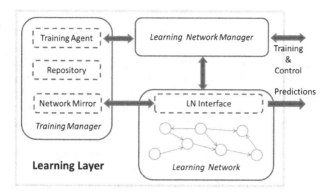

Fig. 1. High-level software architecture of the learning system

3   Learning Layer Architecture

The Learning Layer (LL) is the primary learning system of the RUBICON ecology. It implements a distributed, general purpose learning system where independent learning modules are embedded on sensors and on more powerful devices, e.g. PCs. The high-level goal of distributed learning system is to deliver short-term predictions based on the temporal history of the input signals, including, for instance, the recognition of events ongoing in the monitored environment.

The LL is a complex software system organized into 3 subsystems (see Fig. 1) that are: the *Learning Network* (LN), the *Learning Network Manager* (LNM), and the *Training Manager* (TN). Each subsystem is realized by a variable number of software components that are distributed over an heterogeneous network comprising both sensor and more powerful devices such as robots and PCs.The LL is implemented in NesC [16] for sensors and C and Java for powerful nodes.

The LN realizes an adaptive environmental memory for the RUBICON ecology by means of a distributed learning system where independent learning modules (represented as circles in Fig. 1) reside on the ecology nodes, that are devices with heterogeneous computational capabilities, and cooperate through *Synaptic Connections* (thin arrows in Fig. 1) to perform a distributed neural computation. The single learning module mainly processes local information gathered by the on-board device sensors and integrates this with remote inputs received from other learning modules and delivered through the Synaptic Connection mechanism. The LN subsystem, as a whole, implements the computational learning tasks providing the run-time predictions that serve to the RUBICON ecology to achieve its high-level objectives. Further, it is capable of online refining the learned task based on teaching signals provided by the higher levels of RUBICON. The LN can be trained to tackle any well-defined computational learning task, provided that sufficient supervised learning information is available: hence, it implements a general purpose learning system for RUBICON.

The range of computational learning tasks that LL addresses include event detection, support to adaptive planning and control, localization and movement prediction using signal strength information. All these tasks are characterized by a time-dependent nature are they suit to be modeled as time-series prediction problems. For instance, we expect the LN to be capable of recognizing user activities, such as "cooking" or "cleaning", by learning to predict the corresponding class label in response to input sequences of sensor readings (such as pressure sensors on chairs, accelerometers on kitchenware, infrared presence sensors, etc.). Such readings are provided as inputs to the LN, that is trained to recognize ongoing user activities corresponding to characterizing time-dependent patterns in the sensor data (unknown a priori). The LN also exploits feedback information from the Control and Cognitive layers of RUBICON, which provide the LL with measures of "successfulness" of the actions undertaken to complete a plan, as a result of LN predictions. Such measures can be exploited as feedback signals to refine the corresponding LN predictions. Also, the Cognitive Layer can exploit information on novel user activities ongoing in the ecology (due to its ability of novelty detection) to assemble a training set of examples of sensor measurements corresponding to the novel event, in order to train the LN to recognize the novel activity.

The LNM is a Java software agent hosted on a gateway, that is responsible for the setup and management of the LL. It receives control instructions from the other layers of RUBICON to manage the learning phase and the configuration of the LN. Also, it provides the LN with instructions for setting up and destroying Synaptic Connections, to dynamically attach a learning module to the LN when a new sensor joins the ecology, or to gracefully recover from the loss of a learning module (self-adaptation) consequent to the disconnection of a device from the ecology. Further, it interacts with the TM to control the LN training phases, to incrementally learn a new task or to refine an existing task based on the training information received from the upper RUBICON layers.

The TM is responsible for the learning phases of the LL and for the management, training and self-adaptation of the LN. It receives training information from the LNM to update the modules in the LN. As depicted in Fig. 1, the TM comprises the Training Agent, the Network Mirror and a Repository. The Training Agent is a Java component that manages the activation of the training phases of the LL, by processing the control instructions received from the LNM and by orchestrating the learning in the Network Mirror component through appropriate control messages. The Training Agent receives online learning feedbacks from the upper RUBICON layers, and administers the appropriate refinement signals to the LN. Further, it receives training data and stores it into the Repository; these data are used for the incremental training on novel computational tasks, that are then deployed to the LN, once appropriately learned.

The Network Mirror handles the bulkier learning phases of the LL. In fact, a complete retraining of the distributed learning modules cannot be achieved on resource constrained sensors. To this end, we differentiate between the *online refinement* mechanism, that is used to fine tune the predictions of existing computational tasks (done onboard the sensors), and the *incremental learning* mechanism (performed in the Network Mirror), that is used to learn new computational tasks which are later deployed in the LN. To this aim, the Network Mirror maintains a copy of all the LN modules. Such a mirrored copy of the LN is also useful to ensure LN robustness when a device (and its associated learning module) disappears from the ecology. If the parameters of the missing learning module encode critical knowledge, this is maintained in the Network Mirror, which can clone the module and deploy it to a new available sensor.

4   Learning Layer Implementation in WSN

The LN component on-board of each sensor of the WSN implements the ESN network responsible for the *Forward Computation*, i.e. the operation modality in which the learning system produces its output predictions at each clock tick. This includes the implementation of the necessary functions invoked either locally on the same sensor or remotely on different sensors, and all the data structures required for the embedding of the RC network. In order to allow the communication of learning-related information across the RC modules residing onboard the sensors, a proper mechanism, called *Synaptic Channel*, is implemented at the underlying Communication Layer. In the following, the main aspects concerning the design of Synaptic Channels and Forward Computation modality are described.

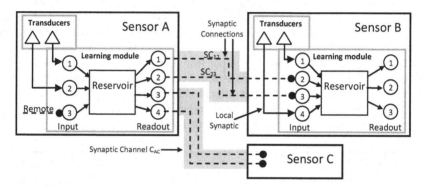

Fig. 2. Synaptic Channel and Synaptic Connection abstractions

Synaptic Channels are a communication mechanism provided by the Communication Layer of RUBICON. They implement the communication abstraction used by the learning modules to perform the distributed neural computation. In particular, a Synaptic Channel C_{AB} is a point-to-point communication channel established between nodes A and B for the transmission of the output information generated by the LL in A to the LL in B (Fig. 2). Synaptic Channels are directional and they are composed of an output end (located on the source node) and an input end (located on the destination node). Synaptic Channels belonging to the same distributed neural computation, are synchronized between them according to a global clock that defines the frequency of the computation. A Synaptic Channel contains one or more Synaptic Connections, each representing the output of a set of neurons in the learning module. Fig. 2 represents two Synaptic Channels each containing 2 Synaptic Connections: C_{AC} that connects node A with node C and contains the output Synaptic Connections 3 and 4 of node A, and C_{AB} that connects node A and node B and contains the output Synaptic Connections 1 and 2 of node A that are bound, respectively, to the input Synaptic Connections 2 and 3 of node B. In the figure are also shown local Synaptic Connections that are used to transfer data read from local transducers to the learning module. All the Synaptic Connections between two sensors are encapsulated into a single Synaptic Channel, thus there exists only one Synaptic Channel for a given pair of sensors. However, a sensor can have several input Synaptic Channels from different sensors, and several output Synaptic Channels to different destinations.

The Forward Computation is the main operational phase of the LL. At each clock tick, computes the predictive output of the RC networks from the input data. The successful completion of a forward computation in a learning module produces a new activation of the internal reservoir units and new values for the readout units. In order to perform a Forward Computation step, each learning module needs all its input data. Such input data can be generated locally to the node (i.e. data produced by local transducers), or it can be received (by means of the input Synaptic Channels) from the learning module residing on remote sensors. The Synaptic Channels implementation ensures that input data is given to the appropriate input units by exploiting a Synaptic Connection mechanism. Let consider the scenario depicted in Fig. 2 and suppose that the learning module on the source A has correctly completed a Forward Computation step, i.e. the output of it RC units has been updated. The LL in sensor A writes the output of units 1 and 2 into the outgoing Synaptic Channel C_{AB} (specifically in the Synaptic Connections SC_{12} and SC_{23}). Similarly, it also writes the output of units 3 and 4 into the Synaptic Channel C_{AC}. In turn, the Synaptic Channels transfer these outputs to the input interface of sensors B and C, respectively. At the next clock tick, the learning module in B acquires the input from the Synaptic Connections SC_{12} and SC_{23} to feed input nodes 2 and 3, and it acquires the data from local transducers to feed input units 1 and 4. Then, the learning module in B progresses with the Forward Computation, producing the outputs on the output units 1,2 and 3, which, in turn, are written in the respective output Synaptic Channels.

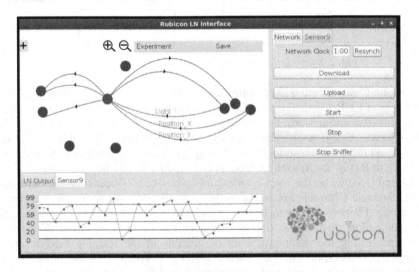

Fig. 3. The main window of the LN interface

The RUBICON LN interface provides a GUI to organize and access the distributed learning system, giving an easy way to adjust the current network settings in order to configure appropriately the LN. By means of the GUI, the user can define new experiments to be tried on the learning network, that are also stored in a database to be retrieved for future executions. Each experiment is defined by a set of devices (sensors) participating in the computation. All devices are seen as opaque objects equipped with an ESN. Inputs to the device can be sent to any of the input neurons of the ESN, while outputs can be read from any neuron in the ESN (input, readout or reservoir). A screen-shot of the typical appearance of an experiment the LN interface is shown in Fig. 3. Here, the structure of an LN network is represented as a graph in which nodes are devices and arcs are Synaptic Connections between them. Each arc is labeled with the identifiers of the neurons connected in the two devices. As connections between modules are usually quite rich, the interface allows to select a single device to focus only on its connections (the red node in the picture). In Fig. 3, we have three incoming Synaptic Connections for *light*, *position$_x$* and *position$_y$* and five outgoing Synaptic Connections to different devices. For each device, the LN interface provides a device manager in which a user can control all the parameters of the ESN loaded on board.

5   Discussion and Conclusions

The RUBICON LL enables all the participants of the ecology to interact and use their past experience to forecast relevant indicators for a changing environment. This feature is of great importance in AmI, since it allows the widespread distribution of learning capabilities in all components of AmI systems, regardless of their processing capacities. The LL is currently available for the TMotes platform [15], and there are short-term plans to publish its code with an open source license. This implementation underwent a deep testing in order to be ready in view of its use in the RUBICON testbeds. Main tests addressed the major functions of LL, namely: the Core Learning Services (that include setup and configuration of the LL and of the LNs), the Local Learning (including testing forward computation and the mechanisms to upload/download learning modules to/from the LN), and the Integration with the Control and Cognitive layers. The LL is now undergoing the experimentation phase of RUBICON, in which it is being tested in living labs and hospitals.

References

1. Baronti, P., et al.: Wireless sensor networks: a survey on the state of the art and the 802.15.4 and ZigBee standards. Computer Communications 30(7), 1655–1695 (2007)
2. Ducatel, K., Bogdanowicz, M., Scapolo, F., Leijiten, J., Burgelman, J.: Scenarios for ambient intelligence in 2010. IST Advisory Group, Tech. Rep. (February 2001)
3. Amato, G., Broxvall, M., Chessa, S., Dragone, M., Gennaro, C., López, R., Maguire, L., Mcginnity, T.M., Micheli, A., Renteria, A., O'Hare, G.P., Pecora, F.: Robotic uBIquitous cOgnitive network. In: Novais, P., Hallenborg, K., Tapia, D.I., Rodríguez, J.M.C. (eds.) Ambient Intelligence - Software and Applications. AISC, vol. 153, pp. 191–195. Springer, Heidelberg (2012)
4. Haykin, S.: Neural Networks: A Comprehensive Foundation, 2nd edn. Prentice Hall PTR (1998)
5. Moustapha, A., Selmic, R.: Wireless sensor network modeling using modified recurrent neural networks: Application to fault detection. IEEE Trans. Instrum. Meas. 57(5), 981–988 (2008)

6. Li, Y., Parker, L.: Detecting and monitoring time-related abnormal events using a wireless sensor network and mobile robot. In: IEEE/RSJ Int. Conf. on Intel. Robots and Systems, pp. 3292–3298 (2008)
7. Nakano, H., Utani, A., Miyauchi, A., Yamamoto, H.: Synchronization-based data gathering scheme using chaotic pulse-coupled neural networks in wireless sensor networks. In: Proceedings of the IJCNN 2008, pp. 1115–1121 (2008)
8. Kulkarni, R., Förster, A., Venayagamoorthy, G.: Computational intelligence in wireless sensor networks: A survey. IEEE Communications Surveys Tutorials 13(1), 68–96 (2011)
9. Lukosevicius, M., Jaeger, H.: Reservoir computing approaches to recurrent neural network training. Computer Science Review 3(3), 127–149 (2009)
10. Verstraeten, D., Schrauwen, B., D'Haene, M., Stroobandt, D.: An experimental unification of reservoir computing methods. Neural Networks 20(3), 391–403 (2007)
11. Jaeger, H., Haas, H.: Harnessing nonlinearity: Predicting chaotic systems and saving energy in wireless communication. Science 304(5667), 78–80 (2004)
12. Gallicchio, C., Micheli, A.: Architectural and markovian factors of echo state networks. Neural Networks 24(5), 440–456 (2011)
13. Gallicchio, C., Micheli, A., Barsocchi, P., Chessa, S.: User movements forecasting by reservoir computing using signal streams produced by mote-class sensors. In: Del Ser, J., Jorswieck, E.A., Miguez, J., Matinmikko, M., Palomar, D.P., Salcedo-Sanz, S., Gil-Lopez, S. (eds.) Mobilight 2011. LNICST, vol. 81, pp. 151–168. Springer, Heidelberg (2012)
14. Bacciu, D., Gallicchio, C., Micheli, A., Chessa, S., Barsocchi, P.: Predicting user movements in heterogeneous indoor environments by reservoir computing. In: Proc. of IJCAI Workshop on Space, Time and Ambient Intelligence (STAMI), pp. 1–6 (2011)
15. Servicios, A.S.Y.: http://www.advanticsys.com
16. Gay, D., et al.: The nesC language: A holistic approach to networked embedded systems. In: Proc. of the ACM SIGPLAN 2003, pp. 1–11. ACM, NY (2003)

Optimizing OSGi Services on Gateways

Iván Bernabé Sánchez, Daniel Díaz-Sánchez, and Mario Muñoz-Organero

Department, University Carlos III of Madrid,
Avda. Universidad 30, 28911, Leganés, Madrid, Spain
{ibernabe,dds,munozm}@it.uc3m.es

Abstract. Currently, the number of devices and services contained in the user's home has considerably grown. Sometimes these devices are provided and managed remotely by service providers by facilitating installation or uninstallation of services. In a typical situation service providers install their services in the user's home gateway from their remote repositories, regardless of the rest of the software installed on it. However, when various service providers work on a given gateway, the number of services and components installed on it increases. Therefore it is possible that this fact may lead to duplication or replacement of components necessary for other installed services, causing performance problems or service interruptions. This paper presents a system for analyzing and automatically optimizing the OSGi components deployed on a home gateway.

1 Introduction

The advancement of technologies in recent years has caused the emergence of new devices with new functionalities and services. These advances offer new possibilities for users in addition have allowed to large companies not only offer new services and but also they are able to remotely manage them in a transparent way to users. For this scenario the service providers use a home gateway device to join a home network to Internet. Service providers (SPs) install on the home gateway the services contracted but when a user contracts various service providers, this device may contain pieces of software from different service providers increasing the possibilities of duplicating of components or software libraries on the home gateway. In this work we present a solution to alleviate the problem on services deployed with technology OSGi [1]. The proposed solution analyzes the services installed on OSGi framework and detects that components may remove of the system without affecting the global system performance. Thus, the system optimizes the software installed by reconfiguring the components installed. Also, we propose to connect various systems to exchange information about its components. If the local system detects that any remote component (installed on a remote system) could improve the performance of local system. It may connect to that system to transfer that component and install it on the local system after of remove the old components. It allows the creating a shared knowledge to help others to improve the performance of the applications installed in their home gateways.

A. van Berlo et al. (Eds.): *Ambient Intelligence – Software & Applications,* AISC 219, pp. 155–162.
DOI: 10.1007/978-3-319-00566-9_20   © Springer International Publishing Switzerland 2013

2   Background

Currently, there are several ways to install applications on the devices. Normally, users first have to get the application, copy it to the target device and finally install it on that device. However, along the years this concept has changed and appeared mechanism that automatically downloads applications between machines and after install them following a client-server architecture model. Some mechanisms aim to go beyond and are able to detect and install the necessary dependencies such as Java Web Start [2]. It is a system for the deployment of Java applications through networks, this system describes the code used and the requirements of a Java application into an XML. Then the user may install the application via the Internet with a special client Java Web Star. However, this system does not support dependency resolution as required by OSGi. Therefore as this work is oriented OSGi platform, this solution is not valid. OSGi specification defines a frame-work inspired on Service-Oriented Architectures (SOA). It allows to manage services with the aim to make up applications based on services. OSGi is usually built on top of Java and each element of the OSGi frame-work is called bundle. A bundle is a software package that is able to provide services to other bundles installed in the OSGi framework. Each bundle is also able to export packages. Thus, OSGi applications are formed by coordinated resources offered from different bundles. OSGi is a key technology for service providers because it allows them to remotely monitor, install and to maintain services deployed on the gateway. There are several ways to deploy OSGi bundles on a platform. One of the most extended is to connect the platforms with centralized reposito-ries of OSGi bundles. From there the users download the desired bundles for use in their applications. However, using bundles from any repository may present a risk to the ap-plications since users may provide bundles intentionally malicious or defective and other users may download them. However, this system is very advantageous to the SP, since them-selves manage their repositories and therefore could rely on bundles hosted on them. However the problem appears when the applications of different SPs have to share the same OSGi framework.

3   Related Work

OSGi frameworks have mechanisms to resolve the dependencies required by a given bundle from other bundles located in the framework. Whether various bundles export the same libraries, the OSGi framework assigns one of them to the bundles requiring it. However, the framework does nothing with duplicated libraries and keeps the loaded bundles in the OSGi framework. There are several works describing solutions to complement OSGi application deployment. However, they do not take into account the problem presented above and they focus to transfer the remote components and create the list of available components in the network. An example of it may be seen in [3] which presents a system to deploy bundles using a P2P network. It relays in Pastry overlay network and provides a scalable system and reliable, however it does not control the configuration of the installed components. [4] shows a solution more advanced than [3], it uses JXTA to rely in P2P protocols and not only is in charge to discovery and download bundles from other systems. But also analyzes the features and requirements of each component and then calculates a score to select which is the best component for installation on the local system. [4] does not take into account the possibility to eliminate duplicated components in the OSGi platform as it is presented in this article. Other solutions also deal with distribution of bundles in OSGi frameworks in different ways like [5] which is focused on increasing the security during bundle distribution from a centralized repository. But it does not offer controls for the management of resources on the framework. A similar reference to the work proposed in this paper with some limitations can be found in [6]. The proposed architecture is able to install the elements necessary when a service provider installs a service for a new device. It uses ontologies to meet with the required services fitting with their requirements. In spite of it selects the best configuration for the new service that will be installed. However, [6] does not consider the components installed on the Gateway from third parties, and it does not optimize its configuration. Briefly summing up, none of the aforementioned approaches try to reduce the re-source consumption or to detect duplicated software components. So, the next sections will present a software architecture to reduce the resources used by the applications on a OSGi platform.

4 Architecture

For clarity, we want to put the reader in context in order to identify the components of our system and their interactions. For that reason, let us to introduce a scenario of application as sketched out in figure 1. The scenario consists of several key elements, service providers (SP), home gateways (HG) and home networks (HN). The user has multitude of services and electronic devices are within theirs home network, however at any given time the user requires a specific service offered by a service provider. The user then contracts the service provider and the provider installs the service in

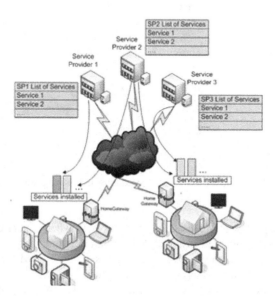

Fig. 1. Scenario of the proposed solution. This scenario shows the most important elements involved where the solution will work.

the user's HG. Later, user sees another interesting service and decides to contract it, the service belongs to another service provider which similarly installs the services on the HG. At this point the user's HG hosting services from different providers, it implies that the HG has installed software developed by independent companies without any contact between them. It gives the possibility that the HG could have similar components installed on the HG and belonging to different SP, because each SP optimize its software for theirs services and they do not take in account the services installed on the GW. It causes to the device higher resource consumption. For this reason, the proposed system identifies the problem and tries to reconfigure the software packages to eliminate the duplication of software on the system.

The proposed architecture adds several services to the HG to control the duplication of SW installed on it. The goal is to eliminate duplicated components in a manner transparent to the user and without affecting the operation of other services installed on the device. This optimization will improve the utilization of resources minimizing the resources used as for instance the memory consumed. As the user may infer, our solution requires nothing but a slight change in the applications installed on the device, just add the services defined in this article which are show in the figure 2. They are four services: Analyze Service (AS) which is responsible for detecting and analyzing bundles installed on the OSGi platform, Storage Service to store and manage information collected by the previous service. Optimizer Service (OS) which uses the information obtained from the other services and is able to obtain the optimal configuration of the system. And finally Distribution Service (DS) is responsible for retrieving information about components located at remote systems provide information about remote packages. Our solution first analyzes the installed bundles, to process the data collected and decides how to reconfigure elements in the OSGi framework to optimize resources. So the solution proposes to model the service such that the developer distributes services and packages in several bundles instead to group these elements in the same bundle.

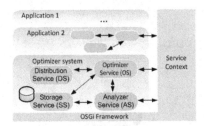

Fig. 2. Schema of blocks. It shows the services implicated in the process to optimize the OSGi framework.

In OSGi the packages are used of several ways, when a service (S1) requires a package, it could be included inside of bundle that contains the service (S1). Whether other service (S2) needs the same package, the service (S2) will have to include the package inside of bundle that contains S2. This behavior does not allow to share similar packages. However, whether the package is packed in a bundle and the bundle exports the package to other bundles of the system. Any bundle will be able to use the package as whether it was contained inside of that bundle. For instance, whether the package is packed inside of bundle P1 and is exported for that other bundles can use it. The service S1 will not need include the package inside of its bundle because may import the package from P1. Same thing happens with S2, S2 will be able to import the package from P1 and could use it. So, S1 and S2 are able to use the same package and due to it is not necessary to load that package on different bundles within the OSGi framework. It is possible because OSGi provide of a class loader for each bundle of the system. That is, each service has a class loader that loads the necessary classes for that service. Figure 3 shows discussed model.

Fig. 3. Structure of services. It shows as to distribute the dependencies of the services.

Our solution is transparent to the services installed by third parties and is also installed on the gateway. The proposed solution consists of two groups. The first group is formed by the services that belong to service providers and the second group is composed by the services presented in this work which are responsible for the optimization of the services contained in the first group. The first service located in the second group is called Analyzer Service (AS) that aims at detecting the bundles installed on the OSGI framework and analyzing each of them in order to find out theirs dependences, and the packages exported by them. The AS requests to the OSGI framework the list of services installed in any moment. Then, for each service, the AS retrieves the manifest.mf file of each of bundle. This file exists on each bundle since it is used to register the required resources used by the bundle as classes, packages, services, etc. For our solution, the AS retrieves the list of exported packages and imported of each bundle, thus the system is able to find out the available resources on the OSGi framework and then be able to calculate the best configuration for its operation. The "Storage Service" (SS) is responsible of management the information collected from the AS. The SS allows to store and consults that information from other installed services on OSGi platform. Mainly the Optimizer Service (OS) uses the SS to carry out the process of optimization. The Storage service's behavior is similar to a data base, it manages the collected information and keeps it updated using the AS and DS. When the AS detects changes on the OSGi framework, it updates the stored local information through the SS. The "Distribution Service" (DS) is in charge of retrieving the information from remote systems. This service provides added value to the solution proposed on this article. Because of it tries to improve the obtained results increasing the available knowledge on the local system. This aim is achieved providing of a distributed communication system to share information of the installed components on the remote systems. Thanks to DS, the local systems not only have more available information but also they are able to transfers components installed between the remote systems. So that, whether a remote bundle is ideal for optimizing the local system, the system will transfer and install that bundle on the local system. To allow the system to share resources with trusted, we may use a mechanism such as that shown in [7]. The Optimizer Service (OS) is in charge to optimize the system with the retrieved information by the services commented above. Within the concept of optimization about this work, the proposed solution tries to detect duplicated software packages used by different bundles duplicated on an OSGi platform. The aim is that packages can be shared with other elements on the OSGi platform to eliminate other bundles. The OS aim is uninstall the unnecessary bundles and keeps the necessary bundles to satisfy the requirements of the other elements. In a real scenario, in general the services installed on the OSGi platform requires of many packages and bundles. In that scenario the packages are distributed among bundles where it is possible that any bundles export a great number of packages (B1) and other bundles export few packages. (B2). In this case, the proposed solution will try to detect whether B1 provides the same resources that B2 export. Then, if it happens the systems uninstalls B2 and the system will continue working. B1 could also include part of resources of B2 instead of including all them. In that case, two things may happen: 1) the system should keep B1 and B2 (because B1 does not meet with the packages provided by B2) or 2) the system tries to detect other bundles (for instance B3) with the aim that they will be able to supply the resources that B1 is not able to meet. It will allow B2 will be uninstalled without affect to the other services installed on the system because the resources provided by B2 are supply by B1 and B3. As a result of this process, the system is consuming less resource because the OSGi platform only has loader the bundles B1 and B3.

5   Operation

This section will explain in detail the sequence of execution of the proposed solution in a runtime environment. The operation is composed of three stages; in the first the proposed solution obtains information on the OSGi framework. Depending of this information, the system will decide if the system has been optimized. If the system must be updated, the OS gets the information from the SS and builds a table with all available information about the available packages offered in each bundle. After the OS creates a list of necessary packages for the OSGi platform meets the requirements of the installed services. At this point, the OS should combine the bundles to find a combination ideal in order to improve the configuration of the system. For this purpose, the solution uses search algorithms which from a root combination are generated combinations which each combination adds a new feature with respect to the parent combination and a different feature regard to the same level nodes and so on. Once the algorithm has generated all possible combinations, it has all possible combinations of the bundles. So also it has all the possible optimal solutions. Although the full deployment of the combinations generated by the algorithm would be the ideal result, because then it will look for among generated combinations the best configuration of platform. In many cases it is not possible due to high consumption of memory and processor used by the algorithm. For this reason we have added three modes of operation that control the generation of combinations: The mode 1 generates all the possible combinations of bundles and later it will take the best combination among all the combinations that meet the requirements. The mode 2 generates all the possible combinations of bundles until that it gets a combination that meet with the defined requirements. The mode 3 uses a heuristic to eliminate the combinations of bundles generated which is exporting the less number of libraries, giving as a result a system with few bundles. It is possible that the algorithm converges and returns no solution. In this case, the system is not updated, because it is not possible to obtain a configuration more optimal than that the current. Once the system has decided which would be the best configuration for the system, there are other factors to take account before install or remove any component. In another paper our previous [8] discussed the factors to consider when it has to make changes in the system.

6   Conclusions

This paper has presented a system to reduce the software resources consumed by OSGi applications deployed on Home Gateways. The solution is focused on reducing the bundles installed on an OSGi system, identifying those bundles that are unnecessary through the analysis and processing of information of the elements installed on the platform. The objective is to obtain the best combination of bundles installed to satisfy the minimum set of elements installed on the OSGi platform. In such a way, the solution will allow that non-professional users can forget about configuring the gateway, except for some preferences that should be defined during the set up. Looking to the users, the big advantage is that the Home Gateway will be able to automatically optimize OSGI applications installed by service providers contracted by the user.

The proposed solution is based on search algorithms which generate combinations of resources with the aim of getting the best combination of bundles to minimizing the number of resources loaded in the OSGi framework. In future works we will design heuristics in order to improve the system efficiency.

Acknowledgments. The research leading to these results has received funding from the ARTEMISA project TIN2009-14378-C02-02 within the Spanish "Plan Nacional de I+D+I", from the European Union's Seventh Framework Programme managed by REA-Research Executive Agency (FP7/2007-2013) under grant agreement n° 286533 and from the HAUS IPT-2011-1049-430000 project funded by the Spanish Ministerio de Economía y Competitividad.

References

[1] OSGi Alliance, OSGi Service Platform Core Specification (2011),
    http://www.osgi.org/download/r4v41/r4.core.pdf
[2] S. microsystems, JavaTM Web Start Overview (2005),
    http://java.sun.com/developer/technicalArticles/WebServices/
    JWS_2/JWS_White_Paper.pdf
[3] Frénot, S., Royon, Y.: Component deployment using a peer-to-peer overlay. Component Deployment, 33–36 (2005)
[4] Schmidt, H., Yip, J.H., Hauck, F.J., Kapitza, R.: Decentralised dynamic code management for OSGi. In: Proceedings of the 6th Workshop on Middleware for Network Eccentric and Mobile Applications, pp. 10–14 (2008)

[5] Parrend, P., Frénot, S.: Supporting the secure deployment of osgi bundles. In: IEEE International Symposium on a World of Wireless, Mobile and Multimedia Networks, WoWMoM 2007, pp. 1–6 (2007)

[6] López de Vergara, J.E., Villagrá, V.A., Fadón, C., González, J.M., Lozano, J.A., Álvarez-Campana, M.: An autonomic approach to offer services in OSGi-based home gate-ways. Computer Communications 31, 3049–3058 (2008)

[7] Bonnaire, X., Rosas, E.: WTR: a reputation metric for distributed hash tables based on a risk and credibility factor. Journal of Computer Science and Technology 24, 844–854 (2009)

[8] Bernabe Sanchez, I., Diaz-Sanchez, D., Organero, M.M.: Optimizing resources on gateways using OSGi. In: 2012 IEEE International Conference on Consumer Electronics (ICCE), January 13-16, pp. 526–527 (2012)

Extremely Small and Incredibly Everywhere

Louise Beltzung Horvath[1], Julia Grillmayr[2], and Tanja Traxler[3]

[1] Vienna University of Technology, Department of Spatial Development,
Infrastructure & Environmental Planning
louise.beltzung@tuwien.ac.at
[2] University of Vienna, Department of Comparative Literature & Department of Philosophy
julia.grillmayr@univie.ac.at
[3] University of Vienna, Department of Physics & Department of Philosophy
tanja.traxler@univie.ac.at

Abstract. Ambient Intelligence technologies call for new space concepts building on an understanding of how humans interrelate with objects. This paper argues for an environmental perspective on the analysis of persuasive technologies. It assumes that dualistic thinking, which recurs to categories such as society/technology or subject/object, has to be questioned. This paper resumes the approach of the PhD project 'Thinking Space' on the spatial dimensions of intelligent technologies. Space concepts from physics, sociology and literature theory form the basis for an empirically informed philosophical approach.

Keywords: Ambient Intelligence, control, smart objects, Ubiquitous Computing, philosophical ecology, smart environment, persuasive technology.

1 Introduction

In bed that night I invented a special drain that would be underneath every pillow in New York, and would connect to the reservoir. Whenever people cried themselves into sleep, the tears would all go to the same place, and in the morning the weatherman could report if the water level of the Reservoir of Tears had gone up or down, and you could know if New York was in heavy boots. [10]

Catastrophe is a rewarding motive when philosophers of technology refer to literature. It means a radical change of the *world* of an individual or a group. Hereby implied is the destruction of the ready-to-handness of everyday practices and objects; therefore, *world* becomes explicit. Jonathan Safran Foer's *Extremely Loud and Incredibly Close* starts with a catastrophe: 9/11 – The nine-year-old Oskar Schell loses his father in the terrorist attacks on the World Trade Center.

This novel serves as leitmotiv to this paper, which explores the theoretical concepts underlying impact evaluations of Ambient Intelligence (AmI) technologies. Thereby, this paper explores how literature comes into play in the analysis of ubiquitous technologies. Literature is acknowledged to hold a source of knowledge of the real world in a more fundamental way, beyond delivering scenarios. As the leading thinker of literary *géocritique* Bertrand Westphal shows, literature makes the imaginary dimension of the real explicit. [37]

A. van Berlo et al. (Eds.): *Ambient Intelligence – Software & Applications,* AISC 219, pp. 163–170.
DOI: 10.1007/978-3-319-00566-9_21   © Springer International Publishing Switzerland 2013

2   Fear and Adoration

What if the water that came out of the shower was treated with a chemical that responded to a combination of things, like your heartbeat, and your body temperature, and your brain waves, so that your skin changed color according to your mood? (...) Everyone could know what everyone else felt, and we could be more careful with each other, (...). [10]

To reflect on the radicality the implementation of technological developments might get in practice, a range of spearing technology analyses in philosophy build upon catastrophes. [29] Martin Heidegger for example may be interpreted as a thinker for whom 'philosophy can only reflect on the catastrophe of technology'. [9, p. 88] His contemporary (and in many ways opponent) Günther Anders was as well driven by the motive of catastrophe: the destructive power of the Atomic Bomb, experienced in Hiroshima and Nagasaki. His writings inherently aim at bringing the technological world, experienced as hostile, back to a human scale. [1] On a more recent account, new information technologies have been analyzed within their contingent geospatial implications by Paul Virilio who points at the failures and accidents as elements of catastrophic imaginary impacts of these technologies on humanity [35].

These approaches, partly coined as cultural pessimism, can be contrasted to technological determinism; generally speaking, thinkers for whom progress is technologically induced [16] and in the extreme leads to utopias of humans harmoniously dwelling in a technologically immersed every-day. The community around the paradigm of Artificial Intelligence has been criticized for the inherent assumption that with technological advancements human and social problems might be solved [see e.g. 11]. Similarly, Oskar confronted with the devastation finds comfort in the idea that new technologies might release him from all kind of problems. His notion of society builds on endless trust upon his fellow people. For him, total transparency is something worth pursuing, because people may grow to be more careful with each other.

Both positions – fear and blind adoration for any technological novelty – contain the wish for some kind of a natural state. Richard Sennett resumed this on the point: 'Fear of Pandora creates a rational climate of dread—but dread can be itself paralyzing, indeed malign. [...] a desire in many of us, that of returning to a way of life or achieving an imaginary future in which we will dwell more simply in nature.' [30, p. 3] Whereas the ones relates to the idea of a perfect harmony with the world mediated by technologies, the others have a return to some primary nature state without technologies in mind. [19, p. 13] When Mark Weiser proclaims Ubiquitous Computing (UbiComp) as a concept of a perfect embedding of machines in our daily environment, which will make 'using a computer as refreshing as taking a walk in the woods' [36, p. 104], he moves within the same logic of those proclaiming scenarios of humans slaved to the will of a machine world. In the context of AmI, utopianism is not the prevalent driving force. Still, 'what might be called specific or single system utopianism still abounds in various beliefs in the technological fix.' [19, pp. 6–7]

It is important to realize that both states can only be reached in theory. The emergence of AmI technologies adds something utterly new to these imaginations. The common point of the different concepts such as UbiComp or AmI is the unobtrusiveness of objects [20, p. 8] as an idea of calmness [36]. This aspect is central because it challenges thinking of technology as *confronting* the human or nature: an immersion of technological artefacts into our environment surpasses the previously experienced, and more evidently challenges the still prevalent thinking [12] within dichotomies.

The persuasiveness of technologies in AmI environments is nothing new per se, there is a long history of analysis to be traced of how artefacts shape perception and actions of humans as social beings, but what changes within AmI is the subtleness this influence takes on. The agency of objects, or the inscription of morality into them, is well explained within the work of the sociologist Bruno Latour, which has substantially contributed to consider not only humans but also things as communicating. [2, p. 17] In persuasive environments though human experience fundamentally changes as we do not encounter technologies but they are present in the background, merged with the built environment and far more dynamically constituted than classical technologies. 'They occupy a radically new position in the realm of human experience. While «classical» technologies are encountered from a configuration of «using» technology, these technologies merge with our environment [...]. Often without us noticing them explicitly, they actively interfere with our lives, in tailor-made ways. Some do so in compelling ways, and others by means of persuasion or seduction; some do so visibly, while others remain largely unnoticed.'[34, p. 232]

3   Controller and Controlled

What about a device that knew everyone you knew? So when an ambulance went down the street, a big sign on the roof could flash DON'T WORRY! DON'T WORRY! if the sick person's device didn't detect the device of someone he knew nearby.[10]

'Total control at your fingertips!'[26] Ambient Intelligence has the explicit objective to be humanist; the driving forces of development should be 'humanistic concerns, not technologically determined ones', and the ability of users to control AmI technologies is considered central in this respect. [20, p. 8] This imposes challenges, (see e.g. [33]) most obviously because the concept of disappearing technologies and the ambition to remain in control are contradictory. [31, p. 12] The concerns framed as questions of privacy [24] lead to an established field within technology studies: the question of control and hence, to which extent the vision of Mark Weiser should be questioned. [28] Hereby, there are several inherent risks in how to address this problem.

The complexity of distributed responsibility has to be acknowledged. The classical example brought forward to contradict any line of argument thinking of *the* designer/developer of a technological invention as *the* responsible, is the physicist Robert Oppenheimer. He had a leading position in the Manhattan Project that produced the first Atomic bomb. Later on he dealt with the outraging consequences his work had had. 'When you see something that is technically sweet, you go ahead and do it and you argue about what to do about it only after you have had your technical success.' [30, p. 2] It is not only that the technologist as the 'inventor' may not control or even predict the consequences of his/her acts. The idea of anything such as *the* controller and *the* controlled as fixed and stable categories is misleading as well. This becomes even more obvious within the complex context of AmI technologies' development.

The relation between controller and controlled may be considered as one of constant interaction and negotiation. With reference to Oppenheimer the philosopher Hannah Arendt for example entrusted the public to debate on technological inventions and hence counterbalance any risks induced by raw technical ambition. 'She had a robust faith that the public could understand the material conditions in which it dwells and that political action could stiffen humankind's will to be master in the house of things, tools, and machines.' [30, p. 4] A more open account to this interplay was formulated by Pierre Bourdieu, who spoke about a field of struggle. 'In short, no one can take advantage of the game, not even those who dominate it, without being taken up and taken in by it.' [6, p. 308][1] His focus on the inscription of social structures in bodily practices and perceptions does deal with the question without defining technologies as such. The intention thereby is not to say that technologies are obsolete to analyzing societal structures and dynamics, but in contrast it shows how from an analytic perspective technologies can be both looked at as artefacts and objects, as well as techniques and practices mediating body-world relations. [32, p. 370] Insofar, as AmI technologies have as a core objective the unobtrusiveness this theoretical perspective opens a way to deal with the questions of control within the appropriate complexity.

The complexity hereby is constituted by different factors: on the one hand AmI technologies are ought not to interfere into the body as human enhancement technologies do, but to shift to the background. This imposes difficulties, as stated before, to properly address from a philosophical perspective the question of control as understanding the agency of objects cannot consist in working with the notion of human agency as bodily rooted agency. On the other hand, AmI technologies are ought to constitute a net which surpasses by its connectivity distances. As such it becomes even more difficult to *name* actors, and thus, to understand *who* controls *whom*. When Michel Foucault explores the society of confinement with reference to the Panopticon, it – at least at first sight – seems possible to *make* someone responsible. There have been thinkers to say that this form of control has been replaced with what Gilles Deleuze coined as societies of control [35, pp. 66–67], in which the *immanence* of control prevails. Thereby, the difference is to be seen in the continuous control. As example serves The Trial of Franz Kafka, in which there is at no point an acquittal or conviction but an ever ongoing postponement. [8] How to within this complexity ever assign anyone responsibility for decisions? '[...], any technology or machine is an expression of a given social form, and is neither its cause nor its effect. Stated differently, machines form part of a given societal assemblage'. [27]

4   Ambient and Environmentalistic

We need enormous pockets, pockets big enough for our families, and our friends, and even the people who aren't on our lists, people we've never met but still want to protect. We need pockets for boroughs and for cities, a pocket that could hold the universe. [10]

The philosopher Peter Sloterdijk devoted his main oeuvre to the idea of humans as creators and sharers of space. This perspective may be called environmentalistic: humans exist inside different spheres, embedded in a multiplicity of relations and through interaction with other human and non-human actors.

---

[1] For Bourdieu technology is a social practice and artefacts form a part of social space.

Extremely Loud and Incredibly Close is a spherical text. Space is always shared and thus requires constant negotiation between its sharers. The novel shows this negotiation down to the most intimate scale of the family and the body. This is made explicit with the example of a couple, which constantly creates 'places' and 'non-places' in their apartment, where they can be together or disappear from another. 'Home is where the most rules are.' [10, p.185] And: 'Home. Where the stuff is.' [10, p.202]

We've wandered in place, our arms outstretched (...), they're marking off distance, everything between us has been a rule to govern our life together, everything a measurement, a marriage of millimetres, of rules, (...).[10]

This ambient perspective is needed for the analysis of AmI impacts. The spatial dimension of these technologies is understated although perfectly recognized within the term of *Ambient* Intelligence. AmI in its application within the urban and the home is 'a new form of production of space' [4, p.3]. Still, there is a tendency to speak about a dissolution of space [23, p. 11] insofar as the virtual part of these technologies is considered detached from its material support. This is inscribed in a long history of neglect of space that has been extensively criticized within the 'spatial turn' [17]. This regards the explicit decision to leave space behind in sociological analyses. This refers also to the assumption of space as ubiquitous to every action, depriving it though of its importance in the context of actions. [7, pp. 21–22]

The difficulty in inquiring the spatial dimensions of AmI lies, we assume, in a lack of adequate concepts to grasp the complexity of these environments. Therefore, sociology of space has since its beginning often referred to and build on space concepts of physics [22] Currently, research on space is a highly relevant topic in physics as one if not the most important challenges of modern theoretical physics consists in the aim to present an over-all-theory of quantum mechanics and general relativity as a whole in a quantum gravity. No satisfactory theory has been found yet despite the great efforts is a lack of data, which are essential to elaborate physical theories. Crucial in this respect is that elements between theory and experimentation like observation and measurement are not well defined within quantum mechanics. Whereas in classical physics object and observer can be considered independent from one another, interactions between them form an incorporate component of the phenomena in quantum mechanics [5]. Hence, also within physics the classical dichotomies of subject/object or nature/culture have to be abandoned [3].

To question dualities in thinking, be it between subject and object, or between nature and culture or technology and society, is our approach to discuss the spatial dimensions of AmI. Subject and object do not pre-exist as such but emerge; Karen Barad refers to this as 'intra-action'. [3] The idea is not to reject dualism as Latour does, by requesting a 'parliament of things' and proclaiming the equality between objects and subjects. In contrast, while acknowledging that reality is more complex than dual pairs of categories, they still form our perception and thus structure our reality. [12, p. 14] Indeed, it is not merely about whether dualisms represent any reality, but what they prevent and whether new metaphors might be of need.

5   Conclusive Remarks: Making Sense

There was (...) a girl with crutches whose cast was signed by a lot of people. I had the weird feeling that if I examined it I would find Dad's writing. Maybe he would have written "Get better soon." Or just his name. [10]

Having lost his *world* Oskar spots clues that may lead him to a new orientation system all around. Every step and action is committed to a hermeneutic of the everyday, which tries to constitute *meaning* out of all this *sense*, distributed into the environment. He collects arbitrary objects in Central Park and draws a map of their locations. By connecting the spots where he found them, he searches for letters and messages. It is a constant translation effort, aiming at putting objects into language to finally understand why all this meaning-less stuff is around. Oskar tries to find the code to decipher the *world*, in which he does not fit anymore. It becomes clear, that by now, his father has had the role of such a code.

This deciphering becomes interesting, when we consider that Ambient Intelligence realises information-intensive environments, which means that information becomes inscribed into the material. The communication of smart objects opens up a tight network of codes. How to think these landscapes of codes is an open question; reading and the relation of text-reader might serve as that purpose. Literary theorists, like the American scholars of Comparative Literature N. Katherine Hayles and Mark Hansen, deal with ambient technologies, [13, 15] since they challenge the relation between the virtual and the material. These debates have a long tradition in literature theory, since a text is considered a material entity (also, when written on a computer screen) on the one hand, but what we refer to as literature is much more than black letters on white support on the other hand. This *more* is not merely imaginary, but opens up in a virtuality which belongs to the real world.

But Oskar does not only *read* objects, he also creates some with a meaning inscribed, as a strategy to regain power over his surroundings. He e.g. hides a message of his father on the answering machine from his mother. But, to cope with the secrecy, he hangs it around her neck in another language:

As for the bracelet Mom wore to the funeral, what I did was I converted Dad's last voice message into Morse code, and I used sky-blue beads for silence, maroon beads for breaks between letters, violet beads for breaks between words (...). [10]

Oskar's bracelet is not smart in the sense of AmI, but his attitude towards objects, makes us reflect upon the relation between material and meaning. If 'materiality emerges from the dynamic interplay between the richness of a physically robust world and human intelligence as it crafts this physicality to create meaning' [14, p. 33], it points to the question how this information materializes and what smart objects do or make us do. Erich Hörl speaks in this context of an 'ecotechnological' order of sense. [18, p. 10] Referring to Heidegger as well as Hayles and Hansen, he stresses the environmental perspective, necessary to fully capture the specific novelty of these technologies. [18, p. 27]. 'Smart dust rather than the Terminator' [15, p. 49] will gain in power and change society; since *sense* is distributed into the environment through small devices, it is accurate to leave the black box of machine-thinking and focus on the objects themselves [18, p. 8].

Acknowledgments. This interdisciplinary PhD project 'Thinking Space' (2012-2015, www.thinkingspace.eu) dealing with the spatial dimension of Ambient Intelligence would not be possible without the DOCteam funding of the Austrian Academy of Sciences. The authors thank Peter-Paul Verbeek, professor of philosophy of technology at the University of Twente for his support as mentor of the project.

References

1. Anders, G.: Die Antiquiertheit des Menschen I. C.H. Beck, Munich (1956)
2. Baecker, D.: Who Qualifies for Communication? A Systems Perspective on Human and Other Possibly Intelligent Beings Taking Part in the Next Society. In: Technikfolgenabschätzung – Theorie und Praxis1, 04/2011 (20), pp. 17–26. ITAS, Eggenstein-Leopoldshafen (2011)
3. Barad, K.: Meeting the Universe Halfway – Quantum Physics and the Entanglement of Matter and Meaning. Duke University Press Books, Durham (2007)
4. Böhlen, M., Frei, H.: Ambient Intelligence in the City. In: Handbook of Ambient Intelligence and Smart Environments, pp. 56–61. Springer, Heidelberg (2010)
5. Bohr, N.: Atomic Physics and Human Knowledge. John Wiley and Sons, New York (1958)
6. Bourdieu, P.: Men and Machines. In: Knorr-Cetina, K., Cicourel, A.V. (eds.) Advances in Social Theory and Methodology. Toward an Integration of Micro- and Macro-sociologies, pp. 304–317. Routledge & Kegan Paul, London (1981)
7. Dangschat, J.S.: Soziale Ungleichheit, gesellschaftlicher Raum und Segregation. In: Dangschat, J.S., Hamedinger, A. (eds.) Lebensstile, soziale Lagen und Siedlungsstrukturen. Forschungs- und Sitzungsberichte der Akademie für Raumforschung und Landesplanung 230, Hannover, pp. 21–50 (2007)
8. Deleuze, G.: Postscript on the Societies of Control. In: Leach, N. (ed.) Rethinking Architecture. A Reader in Cultural Theory, pp. 292–299. Routledge, London (1997)
9. Feenberg, A.: Heidegger and Marcuse: The Catastrophe and Redemption of History. Routledge Taylor & Francis Group, New York, Milton Park (2005)
10. Foer, J.S.: Extremely Loud & Incredibly Close. Penguin Books, London (2005)
11. Geraci, R.M.: Apocalyptic AI: Religion and the Promise of Artificial Intelligence. Journal of the American Academy of Religion 76(1), 138–166 (2008)
12. Gerber, J.: Beyond dualism-the social construction of nature and the natural and social construction of human beings. Progress in Human Geography 21(1), 1–17 (1997)
13. Hansen, M.: Embodying Technesis. Technology Beyond Writing. Studies in Literature & Science. University of Michigan Press, Ann Arbor (2000)
14. Hayles, N.K.: RFID: Human Agency and Meaning in Information-Intensive Environments. Theory, Culture & Society 26(2-3), 47–72 (2009)
15. Hayles, N.K.: Writing Machine (Mediawork Pamphlet). MIT Press, Cambridge (2002)
16. Heilbroner, R.L.: Do Machines Make History? Technology and Culture 8(3), 335–345 (1967)
17. Heuner, U.: Klassische Texte zum Raum. Parodos Verlag, Berlin (2008)
18. Hörl, E.: Die technologische Bedingung. Beiträge zur Beschreibung der technischen Welt, 1st edn. Suhrkamp, Berlin (2011)
19. Ihde, D.: Technology and the Lifeworld. From Garden to Earth. Indiana University Press, Bloomington (1990)
20. IST Advisory Group: Ambient Intelligence: from vision to reality. For participation - in society & business (September 2003)

21. Kastenhofer, K.: Do we need a specific kind of technoscience assessment? Taking the convergence of science and technology seriously. PoiesisPrax 7, 37–54 (2010)
22. Läpple, D.: Essay über den Raum. In: Häußermann, H., Ipsen, D., Krämer-Badoni, T., Läpple, D., Rodenstein, M., Siebel, W. (eds.) Stadt und Raum. Soziologische Analysen, pp. 157–207. Centaurus-Verlagsgesellschaft, Pfaffenweiler (1991)
23. Löw, M.: Raumsoziologie. Suhrkamp, Frankfurt am Main(2001)
24. Maghiros, I., Punie, Y., Delaitre, S., Hert, P., de Gutwirth, S., Moscibroda, A., et al.: Safeguards in a World of Ambient Intelligence. In: Kameas, A.D., Papalexopoulos, D. (eds.) Proceedings of the 2nd International Conference on Intelligent Environments, IE 2006. IET Press, Stevenage (2006)
25. Mattern, F.: Die Informatisierung des Alltags. Springer, Berlin (2007)
26. Philips: Smart Home, http://www.lighting.philips.com/main/subsites/dynalite/dimension/smart_home/lights_camera_action.wpd (February 24, 2013)
27. Poster, M., Savat, D.: Deleuze and New Technology. Edinburgh University Press, Edinburg (2009)
28. Rogers, Y.: Moving on from Weiser's Vision of Calm Computing: Engaging UbiComp Experiences. In: Dourish, P., Friday, A. (eds.) UbiComp 2006. LNCS, vol. 4206, pp. 404–421. Springer, Heidelberg (2006)
29. Rumpala, Y.: Artificial intelligences and political organization: An exploration based on the science fiction work of Iain M. Banks. Technology in Society 34(1), 23–32 (2012)
30. Sennett, R.: The Craftsman. Yale University Press, New Haven (2008)
31. Spiekermann, S., Pallas, F.: Technology paternalism – wider applications of ubiquitous computing. PoiesisPrax 4, 6–18 (2006)
32. Sterne, J.: Bourdieu, technique and technology. Cultural Studies 17, 3–4, 367–389 (2003)
33. Struse, E., Seifert, J., Üllenbeck, S., Rukzio, E., Wolf, C.: PermissionWatcher: Creating User Awareness of Application Permissions in Mobile Systems. In: Paternò, F., de Ruyter, B., Markopoulos, P., Santoro, C., van Loenen, E., Luyten, K. (eds.) AmI 2012. LNCS, vol. 7683, pp. 65–80. Springer, Heidelberg (2012)
34. Verbeek, P.-P.: Ambient Intelligence and Persuasive Technology: The Blurring Boundaries Between Human and Technology. Nanoethics 3, 231–242 (2009)
35. Virilio, P.: The Information Bomb., Verso, London and New York (2005)
36. Weiser, M.: The Computer for the 21st Century. Scientific American 265(3), 94–104 (1991)
37. Westphal, B.: Pour une approche géocritique des textes. SFLGC, Vox Poetica (2005), http://www.vox-poetica.com/sflgc/biblio/gcr.html (February 25, 2013)

Kitchen AS-A-PAL: Exploring Smart Objects as Containers, Surfaces and Actuators

Dipak Surie, Helena Lindgren, and Arslan Qureshi

User Interaction and Knowledge Modelling Group,
Dept. of Computing Science, Umeå University, Sweden
{dipak,helena,mcs10man}@cs.umu.se

Abstract. Technological advancements have taken us closer to the "kitchen of the future" where everyday kitchen activities are seamlessly integrated with smart computing services. While there exist smart kitchen approaches, the explorative nature of the field encourages novel designs. This paper follows the trend by describing the design and development of the Kitchen AS-A-PAL, an infrastructure for facilitating smart kitchen services. Smart objects are the building blocks of Kitchen AS-A-PAL where three types of smart objects namely *Containers*, *Surfaces* and *Actuators* are explored through smart kitchen applications including *interactive cookbook*, *health'n shopping* and *kaffe, god morgon*.

1 Introduction

Kitchen is a place where humans perform important everyday activities like cooking and dining that contribute to their health and wellbeing. Ambient intelligence [1] envisions a future where environments like the kitchen environment could be transformed into a space that offers value added smart computing services, thereby enhancing human experience and improving the quality of the activities performed. Smart objects could be viewed as the building block towards a smart environment, and are defined as *"computationally augmented tangible object with an established purpose that is aware of its operational situations and capable of providing supplementary services without compromising its original appearance and interaction metaphor where supplementary services typically include sharing object's situational awareness and state of use; supporting proactive and reactive information delivery, actuation and state transition"* [2].

The fact that smart objects are viewed as tangible objects can be seen through several prototypes including Teco's Media Cup [3] which is an ordinary coffee cup with additional capabilities like being aware of its temperature, mobility patterns and coffee content level; Energy Aware Kettle that provides energy consumption information and the Smart Medicine Box that presents context-aware services about medicine information and their location [4]; Smart Fridge that detects objects when placed inside it using RFID readers and antennas explored in the pizza lifecycle scenario [5].

There are several smart kitchen services like the Nutrition-Aware Cooking [6] that senses cooking activities and presents nutritional information in real-time for persuading family cooks to make informed decisions; Diet-Aware Dining Table [7] that

A. van Berlo et al. (Eds.): *Ambient Intelligence – Software & Applications*, AISC 219, pp. 171–178.
DOI: 10.1007/978-3-319-00566-9_22 © Springer International Publishing Switzerland 2013

tracks the nutritional intake of persons dining on it and presents such information for inculcating healthy eating habits; Web-based personal multimedia recipe creation tool for using it as memos and for sharing it with others [8]; and automatic analysis of shopping list for inferring nutritional content of the food purchased to support informed shopping list generation in the future [9].

The term tangible objects incorporates certain restriction on the type of objects we consider as smart objects. While an everyday kitchen is not restricted to tangible objects, in this work we try to explore other forms of smart objects that are not tangible but possess specific type of embodiment creating meaning and offering complementary value added services within a smart kitchen context. This paper contributes with the design and development of the Kitchen AS-A-PAL infrastructure and in using it to explore three smart kitchen applications namely interactive cookbook, health'n shopping and kaffe, god morgon.

2   Designing the Kitchen AS-A-PAL

Our design of the Kitchen AS-A-PAL, a smart kitchen is informed by analyzes of four ordinary kitchen environments where human behaviors were observed. In particular, we were interested in human interaction with kitchen objects, both while performing activities and during inactivity. Our observations of the kitchen environments lead to the design of three types of smart objects namely: *containers*, *surfaces* and *actuators*. While actuators are one of the most common types of smart objects that change the state of the world, this paper expands the exploration to include containers and surfaces offering valuable contextual information.

We distinguish between containers and surfaces by defining *containers* as being a placeholder for objects, while a *surface* typically has objects on it while a person is active performing activities. We identified the properties that may influence a human actor's organization of objects, surfaces and containers. The purpose was also to identify contextual information that is valuable in identifying, and defining the properties of containers.

The following factors were influential: *temperature*, *purpose* of an object, the *state* of an object in terms of *time* (e.g., expire dates for milk), *change of location* (may indicate that an object is being used in an activity, e.g., move from a container to a surface) and an object's *relative location to other objects* when included in a set of objects for a certain purpose.

Surfaces typically play the role of scenery for purposeful activity, e.g., eating takes place by the kitchen table. Detecting objects placed on a surface informs the assessment of which activity is being performed. Other important contextual cues are if an object is added to or removed from a smart surface. Such events can be associated to the ongoing and immediate future activities. For instance, by placing the eating plate on the dining table signify that dining activity is about to begin soon. Knowing the beginning and end of an activity is important for offering smart services. The above-mentioned contextual cues would be useful in knowing the beginning and the end of an activity. In addition, observed *relationships between objects with respect to co-location and time* will provide valuable information about the qualitative aspects of the execution of activity.

There are many objects that change their state and thereby offer valuable contextual information. For instance, a coffee machine that can be turned *on/off*, a water heater whose state can be changed from *water absent* to *water present*. While some of the internal state/state change information offers direct contextual cues, there are also contextual cues that can be derived by fusing several states. Smart services can be offered like a coffee machine cannot be turned on without adding water. A change of a state of an object is affected by other changes (or) affect other changes. As a result, simple rules are created as part of our implementation that enables proper functioning of the smart objects.

3   System Architecture

The Kitchen AS-A-PAL system architecture is described in Fig. 1. Networks of PhidgetRFID readers are used in creating 3 smart surfaces wherein a total of 13 RFID readers are used. The smart surfaces read RFID tagged objects using EM4102 protocol with typical read ranges of 7cm. The type of passive RFID tag embedded in objects contribute to read ranges between 3cm and 11cm. The RFID readers are strategically positioned to maximize the object recognition coverage. Standalone RFID readers are used for tracking 5 containers in a semi-automatic manner where implicit tracking is complemented by explicit human actions to improve tracking performance. In total, 36 kitchen objects are passively tagged with each object having multiple tags depending on the object size and shape features. 25 state change sensors are embedded within 6 actuators, 3 containers and 1 surface.

State change sensors offer inference of higher-level contextual cues like *coffee machine has_coffee_powder* using sensor readings over time and logical rules. Such contextual cues are further used as trajectories for activity inference.

Smart Object: Coffee machine (O_1)   **Fluent:** has_coffee_powder {*true, false*}
Action: [add_coffee_powder_O_1, remove_coffee_powder_O_1]
Rule R1: add_coffee_powder_O_1 CAUSE has_coffee_powder {*true*}, has_coffee_powder_time
        {*reset*} IF is_on {*false*}, has_coffee_powder {*false*}, has_coffee_powder_time {*stop*}
Rule R2: remove_coffee_powder_O_1 CAUSE has_coffee_powder {*false*},
        has_coffee_powder_time {*stop*}, has_old_coffee_powder {*false*} IF is_on {*false*},
        has_coffee_powder {*true*}, has_old_coffee_powder {*true*}
Activity: Preparing coffee *(inferred using trajectories)*

Typical sensor types include distance/range (IR reflective sensor 10cm and Sharp distance sensor 2Y0A21), environmental (PhidgetTemperatureSensor IR, relative humidity sensor HM1500LF, precision light sensor), force/pressure (force sensor, FlexiForce 0-100lb resistive force sensor), and motion (motion sensor, 3-axis accelerometer).

Three Toshiba AT100 tablets running Android 3.1 use Phidget21 library to access the sensors. Several I/O boards are used including PhidgetsInterfaceKit 8/8/8 with 6 port hub and dongle sized 2/2/2[1].

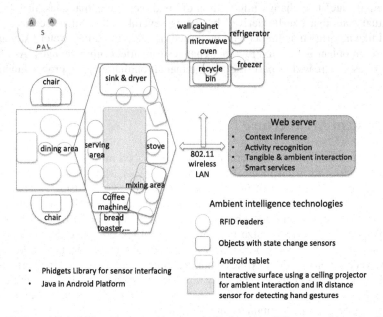

Fig. 1. Kitchen AS-A-PAL system architecture including the different ambient intelligence technologies used with their locations

Raw sensor data are converted to low-level contextual cues in the tablets, while higher-level context inference, activity recognition, smart services, etc. are developed as REST web services. Wireless LAN is used for data communication with the data in JSON (JavaScript Object Notion) data-interchange format. Refer to Fig. 2 for photos of containers, surfaces and actuators that are a part of the Kitchen AS-A-PAL.

4   Applications

Interactive cookbook is an application for the mixing area smart surface of the Kitchen AS-A-PAL that enhances cooking and baking experience by providing context-aware recipe recommendations. The smart services offered by the interactive cookbook are built on the Kitchen AS-A-PAL architecture described in section 3. Objects placed on the mixing area surface are tracked using a network of RFID readers offering tangible interaction [10], while the weight of the ingredients are measured using weight sensors. Human interaction with smart surface using tangible kitchen objects drives the interactive cookbook application. The contextual cues generated when a person places certain object on the smart surface, removes that object, changes

---

[1] For more information about the sensors, I/O boards and Phidgets21 library, visit
http://www.phidgets.com

its location, and adds it to a set of co-located objects enable implicit provision of recipe recommendation service. IR distance sensor supports simple hand gestures in the air above the surface instead of supporting touch-based interaction. A person's hands might be dirty while cooking or baking, and hand gestures in the air remove the need to touch the surface. Information is presented using a projector on the center of the countertop surface (shown in Fig. 1).

Fig. 2. Network of RFID readers and simple state change sensors embedded in containers, surfaces and actuators forming the sensing backbone of the Kitchen AS-A-PAL

When an object is placed on the surface, the interactive cookbook smart services are initiated apart from displaying what the object is. Several recipes are recommended that make use of the objects on the surface and the person can select a recipe of their choice. If there are several objects explicitly registered, then the potential recipe options making use of those ingredients are recommended. Including user profile improves the recommendation capabilities of the interactive cookbook by making it personalized. In Kitchen AS-A-PAL, containers are developed to keep track of the objects they contain like the refrigerator, freezer and wall cabinets. Such contextual information is useful in recommending the appropriate recipes based on available ingredients. The refrigerator smart container described in the next section keeps track of its items, thereby replacing the manual process of checking for ingredients at home by computational services that can provide this information automatically. Time is important contextual information that can be used in recommending appropriate recipes. For instance people are less likely to bake a cake during dinnertime while they would probably not prepare pancakes at 4 p.m.

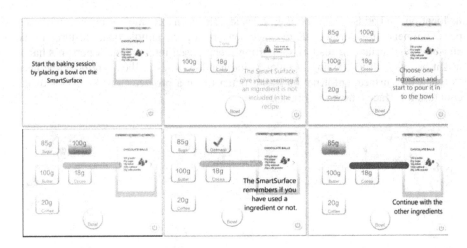

Fig. 3. Six different screen shots from the initial version of the interactive cookbook guiding a person through the activity of preparing chocolate balls

In Fig. 3, the top-left screenshot initiates a person to start the baking session by placing an ingredient for baking on the mixing area smart surface. The top-middle screenshot informs the person about inappropriate ingredients used for baking based on tracking the object on the smart surface. The top-right screenshot initiates the user to select and pour the right ingredient into a mixing bowl. The bottom screenshots from left to right inform about the quantity poured, inform when the sufficient quantity is reached, and remove that ingredient from the ambient display and continue to track the other ingredients used by the person. The smart surface capable of measuring the quantity of ingredients used removes the need for additional measuring tools and to wash them after use.

The smart surface of the mixing area provides the feel and purpose of an ordinary surface with value added computing services as with the definition of smart objects [2]. It is unobtrusive, staying in the background as with Weiser's vision of ubiquitous computing [11] and expects active user participation before recommending recipes using the interactive cookbook. The interactive cookbook enhances recipe recommendations both quantitatively (in terms of the number of recipes available for recommendation) and qualitatively (how useful are they to the current context) in comparison to traditional cookbook.

The **_Health'n shopping_** application for the refrigerator smart container (part of the Kitchen AS-A-PAL) provides smart services for enhancing the occupant's shopping choices for maintaining health and to be on a budget. While it is advisable to buy and eat healthy food, people choices are affected by the unavailability of services that present timely information to persuade them into making healthy shopping choices. The following services are offered by the health'n shopping application:

- **List of Objects Stored.** An occupant of the Kitchen AS-A-PAL can register the items purchased for storing inside the refrigerator and the ones that are desired to be present. The concept is to keep track of the objects stored based on natural human actions.

- **Inventory Level.** Four weight sensors are used on a rack of the refrigerator. The increase or decrease in weight when an object is placed or removed from the refrigerator is used to calculate the object's content and provide information for example to facilitate shopping.
- **Smart Searching.** Objects registered to the refrigerator are passive RFID tagged, where their tags can be considered as electronic product code that provides additional information about the object from its manufacturer. Tangible door magnets representing diary products, expiry date, etc. enable searching for objects inside the fridge and returning the list. Smart searching offers list automatically.
- **Expiry Control.** Items to be expired are presented in an ambient manner by first sending a light signal with no information about which item is to be expired. Tangible door magnet for expiry could be used on the RFID reader for knowing more about the exact items that are to be expired soon.
- **Shopping List with Price Information.** Shopping list can be prepared instantaneously based on the items already available, their inventory level and also their current pricing in the supermarket. While the application is developed as a "proof-of-concept", pricing information from multiple supermarkets in real-time would improve the shopping experience and allow for staying on a budget.
- **Nutritional Value.** The nutritional content of the items stored in the refrigerator in terms of proteins, carbohydrates, fats, minerals, etc.; their consumption rates in terms of how often milk is used in comparison to say mustard; and services to adapt the shopping habits for healthy living are offered.

The *Kaffe, god morgon* application for the smart coffee machine enables occupants of the Kitchen AS-A-PAL to wake-up with some coffee. *Kaffe, god morgon* is an application for smart phone for controlling a coffee machine that can automatically prepare coffee. The application facilitates setting-up alarms, and the snooze feature is unique which persuades a person to move to the coffee machine and take a cup of coffee before the alarm can be disabled or snoozed. The coffee machine sends reminder to the person while setting-up the alarm to fill-up the coffee machine with coffee powder and water. IR distance sensor is used for keeping track of coffee powder availability, humidity sensor for knowing water availability, weight sensor to know the amount of coffee brewed, temperature sensor to know the kettle temperature and a servo motor to turn on and off the coffee machine.

5   Conclusion

This paper has presented the design and development of Kitchen AS-A-PAL using three types of smart objects namely containers, surfaces and actuators with varying properties, tracking technology requirements, and the contextual cues generated. The smart kitchen applications were explored on the different smart object types. The immediate step in the future would be to perform user studies with real users to obtain quantifiable results and evidence of how well the design strategies work in the devised architecture.

Acknowledgments. Elin Sjöström, Viktor Östin, Anna Österlund, Rahel Yitbarek, Selome Tesfatsion, Emil Sjölander, Sofia Papworth, Karl Petersson, and Lisa Sundberg from the Ubicomp 2012 course at Umeå University for developing the applications. This work is partly funded by the Swedish Brain Power.

References

[1] Aarts, E., Encarnacao, J.: True visions: The emergence of ambient intelligence. Springer Publishing Company (2008)

[2] Kawsar, F., Fujinami, K., Nakajima, T.: Prottoy Middleware Platform for Smart Object Systems. International Journal of Smart Home, Special Issue on New Advances and Challenges in Smart Home 2, 1–18 (2008)

[3] Beigl, M., Gellersen, H., Schmidt, A.: Mediacups: experience with design and use of computer-augmented everyday artefacts. Computer Networks 35, 401–409 (2001)

[4] Kawsar, F., Rukzio, E., Kortuem, G.: An explorative comparison of magic lens and personal projection for interacting with smart objects. In: 12th Int. Conf. on Human Computer Interaction with Mobile Devices and Services. ACM, Lisbon (2010)

[5] Schneider, M., Kroner, A.: The smart pizza packing: An application of object memories. In: 4th International Conference on Intelligent Environments (2008)

[6] Chen, J.-H., Chi, P.-Y., Chu, H.-H., Chen, C.-H., Huang, P.: Asmart kitchen for nutrition-aware cooking. IEEE Pervasive Computing 9, 58–65 (2010)

[7] Chang, K.-H., Liu, S.-Y., Chu, H.-H., Hsu, J., Chen, C., Lin, T.-Y., Chen, C.-Y., Huang, P.: The diet-aware dining table: Observing dietary behaviors over a tabletop surface. In: Fishkin, K.P., Schiele, B., Nixon, P., Quigley, A. (eds.) PERVASIVE 2006. LNCS, vol. 3968, pp. 366–382. Springer, Heidelberg (2006)

[8] Silo, I., Mima, N., Frank, I., Ono, T., Weintraub, H.: Making Recipes in the Kitchen of the Future. In: CHI 2004 Human Factors in Computing Systems, pp. 1554–1554. ACM Press (2004)

[9] Mankoff, J., Hsieh, G., Hung, H.C., Nitao, E.: Using low-cost sensing to support nutritional awareness. In: Borriello, G., Holmquist, L.E. (eds.) UbiComp 2002. LNCS, vol. 2498, pp. 371–378. Springer, Heidelberg (2002)

[10] Ishii, H.: The tangible user interface and its evolution. Communications of the ACM 51, 32–36 (2008)

[11] Weiser, M.: The computer for the 21st century. Morgan Kaufmann Publishers (1995)

Ambient Sensorization
for the Furtherance of Sustainability

Fábio Silva, Cesar Analide, Luís Rosa, Gilberto Felgueiras, and Cedric Pimenta

University of Minho, Department of Informatics, Braga, Portugal
{fabiosilva,analide}@di.uminho.pt,
{luisrosalerta,gil.m.fell,cedricpim}@gmail.com

Abstract. Energy efficiency is regarded as an important objective in a world of limited resources. The sustainable use of energy is necessary for the continuity of life styles that do not jeopardize the future. Nevertheless, due to poor information about the impact of human actions on the environment, it is hard to promote and warn for sustainability. This work focuses on the use of ambient intelligence as a mean to constantly revise sustainability indicators in a way they may be used for user awareness and recommendation systems within communities. The approach in this research makes use of sustainable indicators monitored through ambient sensors which enable user accountability concerning their actions inside each environment. Also, it is possible to compare the effect of user actions in the environment, enabling decision making based on such comparison factors.

Keywords: Ambient Intelligence, Sustainability, Energy Efficiency, User Awareness.

1 Introduction

Energy efficiency represents optimal use of energy to satisfy the objectives and needs from users, environments and interactions between them. According to Herring studies [1], over the last 25 years, the increase in the efficiency of domestic appliances has been nullified by the increase of the use of energy consumption devices. Initial results from energy efficiency policies state that small changes in habits can save up to 10% in home energy consumption [2]. On the other hand, sustainability represents the assurance that environments, users and interaction between them can be endured and, as a consequence, the future replication of the current patterns is not compromised.

Both concepts, sustainability and energy efficiency, are not opposed to the use of energy, but they do remind people to be effective on how resources are used and the fact that sustainability concerns the viability of current actions in the present and in the future. Currently, different approaches to measure and assess sustainability are addressed in the literature. Some focus on an economic perspective while others emphasize environmental or social perspectives [3]. On a computer science perspective, although not being able to directly solve the sustainability problem, it can plan and develop solutions to measure and assess sustainability automatically from an environment. This is not due without obtaining information about the environment and its

A. van Berlo et al. (Eds.): *Ambient Intelligence – Software & Applications,* AISC 219, pp. 179–186.
DOI: 10.1007/978-3-319-00566-9_23   © Springer International Publishing Switzerland 2013

users. The scientific research field of Ambient Intelligence provides a wide spectrum of methodologies to obtain such information in a non-intrusive manner.

The types of sensors used in the environment may be divided into categories to better explain their purpose. Generally, an ambient might be divided by sensors and actuators. Sensors monitor the environment and gather data useful for cognitive and reasoning processes [4]. Actuators take action upon the environment, performing actions such as controlling the temperature, the lightning or other appliances. In terms of sensorization, environment sensors can be divided into sensor that monitor environment or sensor that monitor the user and its activities.

This division of sensor classes can also be presented in a different form, taking into consideration the role of the sensor in the environment [5]. In this aspect, sensors might be divided into embedded sensors are installed on objects, context sensors provide information about the environment, or motion sensors.

The work here presented considers the use of these three types of sensors to assess and reason about sustainability and energy efficiency.

The use of indicators for sustainability assessment is a common practice across many researchers. Nevertheless, the definition of a sustainable indicator is sometimes difficult and it may differ from environment to environment. In intelligent buildings, there are proposals to build Key Performance Indicators (KPIs) to monitor sustainability and act as sustainable indicators [6]. It has also been identified that indicators are useful at pointing unsustainable practices but not so accurate nor useful to define and guarantee sustainability [7]. Frameworks to evaluate energy efficiency through sustainability in the literature use similar approaches. The goal of energy efficiency was obtained optimizing sustainable indicators which monitor a set of specific energy sources [8]. Industrial environments are also object of energy efficiency projects. In Heilala et al. [11], an industrial AmI is proposed to optimize energy consumption. The main technique used by the AmI system is based on case based reasoning, comparing the data gathered and processed in the AmI with EUP values to assess and diagnose possible inappropriate energy usages. An intelligent decision support model for the identification of intervention needs and further evaluation of energy saving measures in a building is proposed Doukas et al. [7]. The demonstrated concept shows that it is possible to have an intelligent model to perform energy management on a building, combining aspects like ambient climate conditions, investment rates, fuel, and carbon prices, and, also, past experiences.

2   Sustainability

2.1   Definition

Sustainability is a multidisciplinary concept related with the ability to maintain support and endure something at a certain rate or level. The United Nations have defined this concept as meeting the needs of the present without compromising future generation to meet their own needs. Due to the importance of sustainability, different authors have defined measures to assess and characterize sustainability. A popular consensus is based on 3 different indicators, used to measure the sustainability of a given environment [9]. This approach is based on three different types of indicators, social, economic and environmental, with the specific restriction that until all those values are met, a system cannot be deemed sustainable. From this perspective, sustainability concerns a delicate equilibrium between different indicators, where actions to optimize one indicator might affect anyone of the other two. As a consequence, sustainability planning becomes a hard problem, involving multi-objective optimization techniques, whereas the best solution might not concern the optimization of individual indicators, but rather a compromise between all of them.

2.2 Human Response to Sustainability

It is intended to have users involved and motivated to the sustainability issues, even when some distress may arise from its experience on the environment. Thus, to improve the user's approach to sustainability management platform, it is necessary to reduce this emotional distress. To do so it is necessary to replicate human behavior and emotions, approaching psychological models.

Affective Computing is a computational area that provides techniques for the simulation of emotions, personalities and behaviors, introduced by Piccard. This simulation concludes that cognitive and affective states of humans can reduce the non-determinism of decision making of robots and virtual characters by giving them another level of intelligence [10]. To simulate emotions, models such as the OCC model [11] and the PAD space [12] are common among computational researchers. For personalities, the big five factor model explains how personality is constructed mapping values into five variables: extraversion, agreeableness, conscientiousness, openness, neuroticism, being possible to replicate an human personality [13].

Emotion and personality are intrinsic characteristics of the behavior control, so it is necessary to process the information of an external event. Kazemifard et al. present a model to do that [14]. They separate the information processing into three levels such as reactive level that receives external information triggering a proto-emotion like the associative component of thinking, a reflective level that receives internal information like the unconscious and a routine level, a rule-based component which controls the other levels. This computational model interprets the flow of information from external events to internal change of cognitive and emotional states which might open the possibility to correspond the changes of sustainability indicators to user behavior.

2.3 Sustainable Indicators

The sustainability of a system may be pointed out by a set of indicators, as suggested by many authors, allowing the definition and monitoring of indicators. However, there are common problems with this practice, enumerated in the literature, [8]. The definition of global sustainable indicators, as a means to compare environments, is difficult since environments have different characteristics. Selection and formal definition of indicators is, also, a matter of concern as it has to be agreed by all intervenients and must have a series of properties, in which the indicators express their relevance. Some authors approach this problem characterizing these properties as dimensions, where some indicators are more important in some dimensions than in others, while monitoring the same object. One other problem is the definition of measuring units and metadata. If not defined accordingly, it may be impossible to compare indicators of the same type. Measuring data makes it possible to obtain an indicator which might have a range of optimal values and a range of non-optimal values.

Finally, the presence of indicators to assess sustainability is a common practice. However it does not give any information on how to guarantee or plan sustainability. In fact, indicators only inform about the current status of a system. This work focuses on using sustainable indicators that are built using a common strategy with the same units within the same range of values to facilitate integration with learning and ranking algorithms from ambient intelligence.

3   Sustainability Assessment with Ambient Intelligence

3.1   Sustainability Assessment

As detailed in section 2, indicators are able to detect inefficiencies, but they cannot provide means to guarantee sustainable actions. Therefore, their construction should facilitate their integration on reasoning platforms and algorithms, so they can be used to help achieving sustainability and improving solutions [15]. In this work, the assessment of sustainability focuses on three key dimensions: economic, environmental and social. Within each category, indicators are defined to monitor interest variables inside each category. Indicators are built measuring the positive and negative impact of key variables in the system, and their values are defined as shown in equation (1).

$$\text{Indicator(positive,negative)}=\begin{cases} \dfrac{\text{positive}}{\text{negative}}-1 \rightarrow \text{positive}<=\text{negative} \\ 1-\dfrac{\text{negative}}{\text{positive}} \rightarrow \text{positive}>\text{negative} \end{cases} \quad (1)$$

In this proposal, each indicator as a common scale representing a ratio defined in the interval [-1; 1], where negative values represent unsustainability and positive values respect to sustainability. Moreover, it is possible to aggregate values using simple averaging functions, from a small to a larger perspective. All these indicators are calculated either locally, i.e., in a room basis, or globally, i.e., environment. Thus, even if the environment is considered sustainable, the user may still assess changes in premises with unsustainable values.

Table 1. Sustainability Indicators

		Economic	Environmental	Social
Sample	Positive	Budget	Emissions Avoided	Time Inside
Indicator	Negative	Cost	Emissions	Time Outside

The formal definition of indicators is an active research field. There are different proposals for indicator selection and definition [6], [8]. With the purpose of testing the definition of indicators with the strategy presented in section 2.1sample indicators for each dimension of sustainability were created as demonstrated in table 1. These sample indicators were used to perform experiments in section 4. In order to deliberate about sustainability performance it is needed to rank solutions, by rewarding each solution with a sustainable score. This score can then be used to assess and compare environments inside communities, helping users improve their scores by sharing good behaviors across social network platforms, which presents the users with examples of the best scoring solutions, so they can improve their score.

$$S_{index} = \alpha * I_{economic} + \beta * I_{enverinmental} + \gamma * I_{social} \tag{2}$$
$$\alpha + \beta + \gamma = 1 \wedge 0 < \alpha < 1 \wedge 0 < \beta < 1 \wedge 0 < \gamma < 1$$

The ranking system was designed to take into account the three dimensions of sustainability averaging them with variable weights per dimension as expressed by equation 2. Although the dimension weights should ideally be equal, the expression designed allows the discrimination to account for the promotion of strategies.

3.2   People Help Energy Savings and Sustainability (PHESS)

PHESS is a research project under development at University of Minho which aims to measure the sustainability of environments and its users. The approach focuses on the user and its interactions with environments, assessing their impact in terms of sustainability. The main objective is to build an ambient intelligence platform to promote overall energy efficiency and sustainability.

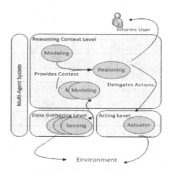

Fig. 1. Multi-Agent System for Deliberation and Sustainable Assurance

Thus, the sustainability assessment used in this work was embedded in a multi-agent system (figure 1) that has, as its primary mission, the management of data and information flow across the community of users, and the promotion of sustainable behaviors. PHESS is built upon a multi-agent system divided into 3 components: data gathering, reasoning and actuating. The data gathering component is composed of agents, sensing agents, responsible for constantly monitoring the environment. All data transformation is made in the reasoning component, as well as the indicators definition and calculation, per environment and user. In this component there are two types of agents: a model agent that models data coming from sensing agents which combines data from sensors to environment and user representations in the agents of this layer. Reasoning agents use agents that model users and environments to recreate simulations combining both where it is possible. It extracts information from both models and combines those models in the simulation engine to test sustainability hypothesis through the values of sustainability indicators. The acting level uses information processed by the reasoning context level to propose and alert users.

4   Results

A number of validations using the platform were designed and implemented in order to prove the practicability and usability of the concepts detailed. As a result, the platform was integrated in simulated, controlled and restricted environments. It was responsible for sensorization and reasoning generating reports and recommendations to users with the aim of reaching better levels of sustainability and energy efficiency.

Table 2. Simulation results for user and environent tracking

Environment 1			
	Social	Economic	Environmental
Kitchen	-0.9011	-0.6859	-0.3263
Bedroom	0.1818	0.9936	-0.3263
Living Room	-0.5294	0.1040	-0.3263
Hall	-0.9690	0.9968	-0.3263
WC	-0.9900	0.9968	-0.3263
Environment 2			
	Social	Economic	Environmental
Kitchen	-0.8889	-0.6231	-0.3263
Bedroom	0.0833	0.9946	-0.3263
Living Room	-0.4849	0.2533	-0.3263
Hall	-0.9690	0.9968	-0.3263
WC	-0.9900	0.9968	-0.3263

For testing purposes, full-fledge environments were defined with a set of rooms commonly found under residential setups. The two environments simulated contained a total of 5 rooms, a bedroom, a living-room, a kitchen, a bathroom and a hall providing connection between all the other rooms. Appliances were defined ranging from lights and computers to ovens and refrigerators with different consumption models in each environment. Consumption of appliances was defined from their active use and explicit power on/off actions from simulated user actions and default consumption models. User actions included movement between rooms and appliance switching on and off. Although, the user performed the same generic action their duration, time and order was different, as well as, the time each user spends inside the environment. Table 2 demonstrates the user report generated by the PHESS system on two users under different environments for one full day.

In order to produce the report an average cost was defined for each room and presence indicators represent the total time with users inside the room versus total time without users in the room. This simulation demonstrates the identification through the use of sustainable indicators, the pattern of sustainability in two environments. Due to the fact that both environments where used by only one user, the social indicator has negative values for most of the room because they are mostly vacant, although the environment itself contains users on it. This results in the difference in the social indicator on a room by room analysis and environment analysis. Environments with more users would help the value of the social indicator if more rooms become occupied at the same time. Regarding the environmental indicator, it is calculated from electrical consumption. As both environments have the same electrical source, they have the same value for this indicator. This indicator may vary if the electricity provider user more or less green sources. The economic indicator is directly connected with the price of electrical consumption and the amount available.

Table 3. Taking Advantage of Information and Knowledge

Environment 1	Social	Economic	Environmental
User 1	-0.004	0.3241	-0.586
User 2	0.5	0.927	-0.586
Environment 2	**Social**	**Economic**	**Environmental**
User 1	-0.004	0.2016	-0.586
User 2	0.5	0.827	-0.586

Using PHESS system and user models created it is also possible to recreate one user behavior in other environments, provided environments are compatible as it is the case. Re-running the simulation using stored user behaviors showcases the potential to compare the effect on specific user behaviors in different environments and how environment and sustainability indicators are affected. In table 3, it is demonstrated that user 2 would perform better environment 1 while user 1 would not be as sustainable in environment 2, thus making environment 2 considered less sustainable for that group of users. Moreover, information collected efficiently aggregates different indicators into each dimension of sustainability as defined in section 3 providing comprehensible information to the user.

5   Conclusion

The use of AmI technology is a valid effort to track sustainability on a real time basis, enabling user accountability for their action inside environments. The results provided in this work demonstrate that it is possible to use a sustainability assessment to directly compare the sustainability performance of both users and environments. Furthermore, the use of AmI techniques enables users and environments profiling testing if a better distribution of users per environments results in better sustainable indicators.

In the future, there is the need to integrate more reasoning context with current and past indicators to create recommendations on the platform. Moreover, the definition of more sustainability indicators is necessary to test the robustness of the sustainability assessment engine. Also, the deployment of the PHESS system is scheduled in order to test the findings found with real environments and users.

Acknowledgements. This work is funded by National Funds through the FCT - Fundação para a Ciência e a Tecnologia (Portuguese Foundation for Science and Technology) within projects PEst-OE/EEI/UI0752/2011 and PTDC/EEI-SII/1386/2012. It is also supported by a doctoral grant, SFRH/BD/78713/2011, issued by the Fundação da Ciência e Tecnologia (FCT) in Portugal.

References

1. Herring, H.: Energy efficiency—a critical view. Energy 31(1), 10–20 (2006)
2. Chetty, M., Tran, D.: Getting to Green: Understanding Resource Consumption in the Home. In: Proceedings of the 10th International Conference on Ubiquitous Computing, pp. 242–251 (2008)
3. Singh, R., Murty, H., Gupta, S., Dikshit, A.: An overview of sustainability assessment methodologies. Ecological Indicators 9(2), 189–212 (2009)
4. Aztiria, A., Izaguirre, A., Augusto, J.C.: Learning patterns in ambient intelligence environments: a survey. Artif. Intell. Rev. 34(1), 35–51 (2010)
5. Aztiria, A., Augusto, J.C., Basagoiti, R., Izaguirre, A., Cook, D.J.: Discovering frequent user-environment interactions in intelligent environments. Personal and Ubiquitous Computing 16(1), 91–103 (2012)
6. Al-Waer, H., Clements-Croome, D.J.: Key performance indicators (KPIs) and priority setting in using the multi-attribute approach for assessing sustainable intelligent buildings. Building and Environment 45(4), 799–807 (2009)
7. Lyon, A., Dahl: Achievements and gaps in indicators for sustainability. Ecological Indicators 17(0), 14–19 (2012)
8. Afgan, N.H., Carvalho, M.G., Hovanov, N.V.: Energy system assessment with sustainability indicators. Energy Policy 28(9), 603–612 (2000)
9. Todorov, V., Marinova, D.: Modelling sustainability. Mathematics and Computers in Simulation 81(7), 1397–1408 (2011)
10. Picard, R.W.: Affective Computing. The MIT Press, Cambridge (1997)
11. Ortony, A., Clore, G., Collins, A.: The cognitive structure of emotions (1990)
12. Gebhard, P.: ALMA: a layered model of affect. In: Proceedings of the Fourth International Joint Conference on Autonomous Agents and Multiagent Systems, pp. 29–36 (2005)
13. McCrae, R.R., John, O.P.: An introduction to the five-factor model and its applications. Journal of personality 60(2), 175–215 (1992)
14. Kazemifard, M., Ghasem-Aghaee, N., Ören, T.I.: Emotive and cognitive simulations by agents: Roles of three levels of information processing. Cognitive Systems Research 13(1), 24–38 (2012)
15. Silva, F., Cuevas, D., Analide, C., Neves, J., Marques, J.: Sensorization and Intelligent Systems in Energetic Sustainable Environments. In: Fortino, G., Badica, C., Malgeri, M., Unland, R. (eds.) Intelligent Distributed Computing VI. SCI, vol. 446, pp. 199–204. Springer, Heidelberg (2012)

Distributed Intelligent Rule-Based Wireless Sensor Network Architecture

Antonio Cubero Fernández, José María Castillo Secilla,
José Manuel Palomares Muñoz, Joaquín Olivares Bueno, and Fernando León García

Computer Architecture, Electronics, and Electronic Technology
University of Cordoba, Cordoba, Spain
{acuberof,fndoleon}@gmail.com,
{jmcastillo,jmpalomares,olivares}@uco.es

Abstract. This paper describes the design of a new system architecture for monitoring and controlling purposes of a group of sensors and actuators within a wireless sensor network (WSN). This system can manage an undefined amount of clustered networks. The proposed system architecture enables Internet communications to reach the WSN in a highly efficient way. This structure reduces the bottleneck of the Internet/WSN bridge and the amount of messages inside the WSN when an Internet request arrives. Besides, each individual WSN implements an Intelligent Rule-Based System Automation (IRBSA) that performs the automation of the behaviour of the network motes according to the previously included rules. These rules describe the actions that are executed when all the conditions of that rule are met. Opposite to traditional approaches, IRBSA is placed in the WSN Header Mote rather than in the Internet server or in every mote.

1 Introduction

Nowadays, wireless comunications are continuously growing both in industry and in domestic environments. It is easy to make an effective and secure wireless sensor network (WSN) compared with wired networks. Currently, many WSN are opening their capabilities to the Internet, offering the data acquired by the sensors included within the WSN to the World Wide Web. Many efforts have been made by researchers in this field, paying special interest in the conception of the Internet of Things (IoT) [7]. This approach tries to interconnect any wired or wireless device, even with different types of hardware and physical network modems. The first step to get an Internet–conected WSN is to have a front–end server (usually a Web server) acting as a bridge between the Internet and the WSN. This structure is very suitable for any WSN with a small number of nodes (usually, called motes) and with very few requests from the Internet side. Each time an Internet request arrives at the Web server, the computer translates the Internet query to the WSN, with a message to a specific device address, requesting the value of an input pin, or assigning a value to an output pin, etc. However, this structure collapses when the WSN enlarges or when the amount of Internet requests rise.

Next stage in the WSN development roadmap is to build fully distributed systems, in which system workload is shared among multiple devices. Besides, fully distributed WSNs allow to introduce fault–tolerance. If any component fails, it does not affect the

A. van Berlo et al. (Eds.): *Ambient Intelligence – Software & Applications*, AISC 219, pp. 187–194.
DOI: 10.1007/978-3-319-00566-9_24   © Springer International Publishing Switzerland 2013

whole system. Some mechanisms are included in the WSN infraestructure to change the role of an idle mote to handle the tasks of the faulty one. Some other benefits may be obtained from fully distributed WSN, for example, these systems allow to enlarge dinamically the amount of motes in an easy way. Thus, this makes the systems to be stronger and more powerful.

The main objective of this project is to design a distributed and intelligent system to manage multiple WSN. Each WSN acts independently, although, some of them are joined to work cooperatively forming larger WSN. Each WSN has its own Intelligent Rule–Based system being in charge of automation of the motes of that very WSN. A communication platform will be created for the interconnection of the WSN. Each WSN has a WSN Header Mote (WHM), which acts as a Coordinator device of that WSN. Each WHM is able to communicate directly with the communication platform. This communication platform makes possible to handle the data extracted from every WSN. However, the communication platform also offers a entrance gate for the Internet to the whole system, to request or update any data of any mote in any of the WSNs. The communication platform is optimized using databases to store the data about WSNs. This fact allows much faster readings of the data of the sensors linked to a mote, without the necessity to ask directly the involved mote to obtain the data. This view of the database will be replicated in each WHM to make every WSN completely self–sufficient.

This article has been divided in several sections. Section 2 makes a short scientific revision about current articles related to the proposed system. The proposed System Architecture is described in Section 3. The Intelligent Rule–Based System Automation is stated in Section 4. Some prototypes are described in Section 5. The conclusions of this work and some future work are presented in Section 6.

2 Scientific Review

Many researchers have used WSN to monitor and control enviroments. Some authors have designed very interesting systems in order to communicate WSN with the Internet while collecting large data from sensors. André et al. [3] proposed a model for monitoring WSNs based on a REST Web service and XML messages to provide a mobile ubiquitous approach for WSN monitoring. Data collected from WSN are stored in a database, although every request from the Internet produces a WSN message to the end–node which has the requested sensor.

Serdaroglu and Baydere [6] studied a proxy–based and gateway–based system. They defined an hybrid approach which combines the advantages to interconnect WSN and the IP networks. The proposed approach is used to build a web server for WSNs. The goal of the study is to reduce memory footprint of the overall system and use possibly small amount of resources of a WSN node implementing in a middleware rather than a full conventional stack or a ready solution.

Previous approaches make the data analysis in devices that are not inside the WSN, either in remote nodes in the Internet or in the Internet proxy or gateway. For these models, WSN are non–intelligent and any Ambient Intelligence strategy must be addressed by elements outside the WSN.

The opposite approach is to include an intelligent data processing in every mote, making the WSN not be a mere network infraestructure, but an intelligent one. Labraoui *et al.* [5] proposed a new scheme for data aggregation in large–scale WSN. However, authors focused mainly in fault–tolerance in clustered–based WSN.

Tapia *et al.* [8] proposed an innovative platform that addressed the requirements of Ambient Intelligence paradigm, such as context–awareness and ubiquitous communication, allowing the use of heterogeneous WSNs and taking advantage of the use of intelligent agents directly embedded on wireless nodes. This article described the integration of the HERA (Hardware-Embedded Reactive Agents) platform into FUSION@ (Flexible and User Services Oriented Multi–agent Architecture) [1,9]. This way, through the integration of HERA and FUSION@, there was no difference between a software and a hardware agent.

Another issue to be analyzed are the Rule–Based Systems in WSN. Rule–Based system are used as a way to store and manipulate knowledge to interpret information in a useful way. They are often used in artificial intelligence applications and research. They consists of a rule–base (permanent data), an inference engine (process), and a workspace or working memory (temporary data). Knowledge is stored as rules in the rule–base (also known as the knowledge base).

Dressler *et al.* [2] designed a Rule–Based Sensor Network (RSN) that mimics the cellular signaling communication. That model has data–centric communications and the rule–based programming scheme describes specific actions after the reception of specific data fragments for simple local behaviour control. RSN is able to process sensor data and to perform network–centric actuation according to a given set of rules. In particular, this system is able to perform collaborative sensing and processing in SANETs with purely local rule–based programs.

All these systems are suitable for the required objectives. However, they demand computationally powerful platforms to be executed, and this work tries to minimize the computational requirements of the WSN nodes, while providing similar results in terms of performance.

3   System Architecture

In this section, the system architecture is described. Traditional WSN (Fig.1) are compounded by an interface, an internal system application, and an access point to the WSN. The interface allows the user to input the requests and to receive the results. Internal system is usually composed by an application that performs the automation and controls all devices and a warehouse storing data. Finally, the WSN is formed by some motes, responsible for gathering data from the environment. This structure is not very scalable and it can be easily overloaded. Therefore, every request from the interface (external orders) would be translated into, at least, one internal message in the WSN, even though of the requested value has not changed. Thus, large amount of requests arriving from the interface would degrade the wireless medium. It is obvious that this structure lacks generality and a different solution is requiered. For that reason, a fully distributed system, called DIRB–WSN (Distributed Intelligent Rule-Based – Wireless

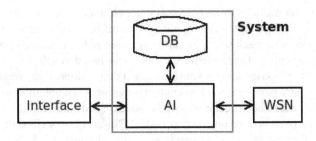

Fig. 1. Typical Internet–connected WSN system architecture

Sensor Network), is proposed (Fig. 2). To achieve this objective, the system is divided into two different modules: *Communication_AI* and *WSN_AI*.

Communication_AI manages input connections from the Internet and the subsequent responses. This module reduces the amount of messages to WSN using *federated tables* for each WSN. The Database Managemente System (DBMS) is in charge of synchronizing the data between database copies allocated in different devices. When a reading sensor query is requested, *Communication_AI* looks for the data in its own copy of the federated table, so that, no message is sent to the WSN. This simple, but very effective, mechanism simplifies the working process of the *Communication_AI*. On the other hand, if a writing or update action is requested by the user, *Communication_AI* sends a message to the *WSN_AI* requesting the modification of the data for a given mote. *WSN_AI* modifies its copy of the federated table (therefore, the *Communication_AI* federated table is also modified by the DBMS) and a message is sent to the wireless mote to modify or update an output value. *Communication_AI* also stores a log for each WSN.

WSN_AI is placed in the Header Mote of every WSN (WHM). This mote is usually the coordinator node of the WSN and it also has another network interface which is able to connect to the *Communication_AI* node. *WSN_AI* is in charge of analyzing and processing the input packets coming from *Communication_AI* and from the inner WSN. It also runs IRBSA (Section 4).

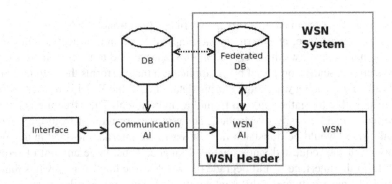

Fig. 2. Proposed Internet–conected DIRB–WSN system architecture

4   Intelligent Rule-Based System Automation

The second part of the project consists of a soft real–time fully automation of the system. For the sake of simplicity, the WSN is assumed to have some kind of time–synchronization protocol, therefore, the time/date of all the motes in the WSN is coherent. The Intelligent Rule–based System Automation (IRBSA) allows to program different behaviours by checking some parameters of the system. IRBSA is composed of a set of rules to be executed locally within each WSN.

Every IRBSA rule have two different parts: *antecedents* and *consequents*. An *antecedent* holds the conditions that must be true for the rule to be executed. Each *antecedent* is composed of the following elements:

- *Active*: it indicates whether an antecedent has to be examined.
- *Date*: date after which the antecedent is evaluated.
- *Period*: time interval for which the antecedent is enabled.
- *Attempts*: maximum number of attempts that the antecedent will be analized. An attempt expires when *period* ends and one or more devices are not still ready.
- *Cycle*: it indicates whether an antecedent date has to be rescheduled with a new future date.
- *Rule*: it states a dependency between the antecedent and one or more rules.
    - These rules must have been activated before the antecedent is checked.
- *Device*: it states a dependency between the antecedent and one or more devices (or other physical constraints).
    - This element imposes a condition that a device must fulfill to activate the rule. For example, the pin value of a device must be greater than another value.

On the other hand, *consequent* indicates an action to be executed in the system. This action ranges from a simple change of the state of an actuator to the reprogramming of a mote, and even to modify the IRBSA. There are two different *consequent* types: *normal consequent* and *error consequent*. Once IRBSA checks an *antecedent*, every rule and device dependencies are analyzed. If an *antecedent* of a given rule is not provided, the *error consequents* associated to the rule are executed. Whereas, if all the *antecedents* are met, *normal consequents* are executed. Besides, if any of the *antecendents* of the rule are cyclical, new time instances of all the *antecedents* are rescheduled. This fact assures that recent values are used for the *antecedents* before the *deadline* of the rule. Figure 3 shows the activity diagram of IRBSA.

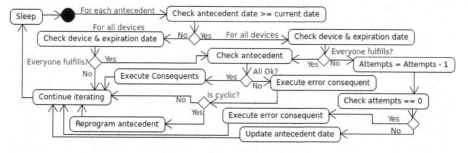

Fig. 3. IRBSA Activity Diagram

Fig. 4. Mote prototype

5   Results

The proposed DIRB–WSN system is a complete functional system, composed of a Internet server with the *Communication_AI* and some WSN systems headered by a *WHM* managing a network of motes. A prototype has been deployed using XBee modules (Zigbee networking) placed on Arduino boards to create the WSN. Figure 4 shows a mote prototype, with a humidity/temperature sensor, a movement sensor, some buttons and switches, and a servo–motor connected to an Arduino FIO board with an XBee modem.

This model has not been compared to any other previous proposals, due to time restrictions. However, many tests have been carried out with different workload and Internet requests. All the tests have provided similar results in terms of number of messages within the WSN. A test scenario was deployed: the WSN was composed of 3 XBee devices, two mote prototypes and a WHM. The WHM was a notebook computer with an Xbee modem and a WiFi port running the *WSN_AI* and a copy of the *federated database*. This WHM was interconnected to a Web server computer. This Web server was running also the *Communication_AI* and the main *federated database*. The experimental test was to simulate 100 Internet queries (50 of them, sensor reading requests and 50 of them sensor updating messages) to the Web server for 20 seconds. One of the mote prototypes obtained the temperature from the sensor once every second. Therefore, the number of messages in the WSN was 70: 20 of them starting from the prototype mote due to the sampling of the temperature sensor and 50 of the messages were because of the modification requests from the Internet side. As the number of messages was relatively low, no collapsing of the WSN was provoked and there were no loss of messages. On the other side, using a standard structure (as described in Fig. 1) would provided a minimum amount of 100 WSN messages (one WSN packet for each Internet request).

Regarding IRBSA, results show that it is a highly efficient system with a quick response to any stimulus, taking less than one second to provide a response in the worst case. This module is able to reduce the amount of messages going outside the WSN to the Internet, where the Ambient Intelligence module could be placed. A simple experimental test was evaluated, by including a rule in the IRBSA in which several sensor values of the prototype were taken into account to activate the servo–motor. For a 60 seconds experiment, the proposed system was able to reduce the number of Internet packets to zero. Meanwhile, for the same experiment, the standard structure with the Intelligent module as a remote application in the Internet side, produced more than 60 Internet messages (more than one each second). Besides, due to WSN collisions and Web server delays, the execution of rules in the standard structure experiment were delayed some seconds, most times.

6   Conclusions and Future Work

This system allows to manage multiple WSN Systems efficiently. The proposed infraestructure and the IRBSA make possible to automate these WSN Systems, which can be managed from a single Internet–conected machine. IRBSA is able to provide some kind of intelligence, keeping very low the complexity of the end–motes without provoking a bottleneck in the Internet gateway connection.

The usage of *federated tables* in distributed databases is a smart mechanism to disseminate information very efficiently. The DBMS is responsible for the synchronization, thus, the code of the *Communication_AI* is simplified as there is no need of checking for concurrent accesses. Furthermore, the data synchronization messages between databases may rely on a different data network, and thus, the amount of messages inside the WSN for reading requests from the Internet can be reduced drastically.

Related to the future work, it is very interesting to include fuzzy logic control (FLC) into IRBSA, in order to evaluate the rules taking both approximated *antecedents* and *consequents* rather than fixed and exact ones. FLC provides a simple way to arrive at a definite conclusion based upon vague, ambiguous, imprecise, noisy, or missing input information [4].

It is planned to make more experiments and to compare the proposed DIRB–WSN with previous proposals. The experiments will be real scenarios with different deployed WSN.

References

1. Corchado, J., Bajo, J., Tapia, D., Abraham, A.: Using heterogeneous wireless sensor networks in a telemonitoring system for healthcare. IEEE Transactions on Information Technology in Biomedicine 14(2), 234–240 (2010), doi:10.1109/TITB.2009.2034369
2. Dressler, F., Dietrich, I., German, R., Krüger, B.: A rule-based system for programming self-organized sensor and actor networks. Computer Networks 53(10), 1737–1750 (2009), doi:10.1016/j.comnet.2008.09.007
3. Elias, A., Rodrigues, J., Oliveira, L., Zarpelão, B.: A ubiquitous model for wireless sensor networks monitoring. In: Proceedings - 6th International Conference on Innovative Mobile and Internet Services in Ubiquitous Computing, IMIS 2012, pp. 835–839 (2012)
4. Feng, G.: A survey on analysis and design of model-based fuzzy control systems. IEEE Transactions on Fuzzy Systems 14(5), 676–697 (2006), doi:10.1109/TFUZZ.2006.883415

5. Labraoui, N., Gueroui, M., Aliouat, M., Petit, J.: Rahim: Robust adaptive approach based on hierarchical monitoring providing trust aggregation for wireless sensor networks. Journal of Universal Computer Science 17(11), 1550–1571 (2011),
   http://www.jucs.org/jucs_17_11/rahim_robust_adaptive_approach
6. Serdaroglu, K., Baydere, S.: Seamless interconnection of wsn and internet. In: 2012 20th International Conference on Software, Telecommunications and Computer Networks, SoftCOM 2012, pp. 1–6 (2012)
7. Tan, L., Wang, N.: Future internet: The internet of things. In: Proceedings of the 2010 3rd International Conference on Advanced Computer Theory and Engineering, ICACTE 2010, vol. 5, pp. 5376–5380 (2010)
8. Tapia, D.I., Fraile, J.A., Rodríguez, S., Alonso, R.S., Corchado, J.M.: Integrating hardware agents into an enhanced multi-agent architecture for ambient intelligence systems. Information Sciences 222, 47–65 (2013), doi:10.1016/j.ins.2011.05.002
9. Tapia, D.I., Rodríguez, S., Corchado, J.M.: In: Hassanien, A.E., Abawajy, J.H., Abraham, A., Hagras, H. (eds.) Pervasive Computing, Computer Communications and Networks, pp. 181–199. Springer, London (2010), doi:10.1007/978-1-84882-599-4_9

Face Detection in Intelligent Ambiences with Colored Illumination

Christina Katsimerou, Judith A. Redi, and Ingrid Heynderickx

Department of Intelligent Systems
TU Delft
Delft, The Netherlands

Abstract. Human face detection is an essential step in the creation of intelligent lighting ambiences, but the constantly changing multi-color illumination makes reliable face detection more challenging. Therefore, we introduce a new face detection and localization algorithm, which retains a high performance under various indoor illumination conditions. The method is based on the creation of a robust skin mask, using general color constancy techniques, and the application of the Viola-Jones face detector on the candidate face areas. Extensive experiments, using a challenging state-of-the-art database and a new one with a wider variation in colored illumination and cluttered background, show a significantly better performance for the newly proposed algorithm than for the most widely used face detection algorithms.

Keywords: intelligent ambiences, adaptive lighting, face detection, skin segmentation, color constancy.

1 Introduction

Intelligent ambiences have embedded the advances of human-technology interaction in arrays of smart sensors and actuators, in order to achieve a natural unobtrusive communication with the user [1]. A specific application of an intelligent ambience we are interested in, aims at building an automatic system which assesses the mood of a person in a room from videos and responds adaptively to it with (multi-color) light settings, which are believed to improve the well-being of the room's occupant [18]. The first and most essential step towards such a system is automatic face detection from the video input, since the face conveys highly relevant emotional information. Therefore, our research focuses on a robust face detector that can handle the challenges of head pose variation, colored illumination, cluttered background and low resolution.

Ambiences with colored artificial light present multiple challenges for face detection. First, these ambiences are created with different light sources positioned in various locations in the room. As a consequence, light on the face is never uniform, but rather unevenly distributed, producing reflections or shades. In addition, colored light sources often alter the skin color captured in the images significantly. Finally, certain ambiences are composed with low intensity illumination, making the faces hard to distinguish.

A. van Berlo et al. (Eds.): *Ambient Intelligence – Software & Applications,* AISC 219, pp. 195–204.
DOI: 10.1007/978-3-319-00566-9_25   © Springer International Publishing Switzerland 2013

A number of robust face detection algorithms have been proposed in the literature, with the most recent ones reviewed in [24]. The main concept of the latest appearance-based approaches is that they collect a large number of positive and negative samples (face and non-face patches) from images, extract features from the intensity component of these patches and feed them as input to a classifier, which is trained to distinguish face from non-face patches. Among these approaches, the Viola-Jones (VJ) algorithm [23] has typically been preferred for detecting upright frontal faces, due to its simplicity and effectiveness.

A general shortcoming of the appearance-based methods is that a high detection rate may result in a large number of false positives, namely non-face patches that are incorrectly recognized as faces. Especially a complex background increases the chance of misclassification. To reduce false positives, complementary information, like skin color, was used in [7, 12], however, without a light compensation technique. As a result, these algorithms missed skin regions, and as such, yielded a lower detection rate than the basic VJ detector. A skin color detector was used in [8] after the Viola-Jones module, in order to filter the correct detections. In this way, the authors achieved a low false positive rate, but did not improve the detection rate.

This study proposes a refined, computationally inexpensive and human inspired face detection method from color images. The proposed algorithm combines robust skin segmentation (applied on the color-corrected image with an optimal color constancy technique) with the publicly available VJ face detection framework [11].The skin segmentation module can successfully detect skin pixels under colored, multidirectional and uneven lighting. The uniform background produced around the skin areas after segmentation can outbalance the clutter and improve the final detection rate, while the localized search of VJ only on the face-candidate regions reduces the false positives significantly. Due to a lack of suitable face databases for evaluating our face detection algorithm under ambient colored illumination, we introduce a new challenging face database, which we named "CI" for Colored Illumination.

2   Overview of the Refined Face Detection Algorithm

In order to choose a suitable face detection algorithm under colored ambience light conditions, we tested the VJ on a subset of the state-of-the-art PIE face database [20]. Preliminary results on 100 images of frontal faces under blue-flash illumination indicated that the VJ algorithm could not handle the colored illumination optimally (10% missed faces, 8% false positives). Therefore, we propose a refined approach, which enhances the basic VJ with a pre-processing step that robustly detects skin regions (Fig. 1).

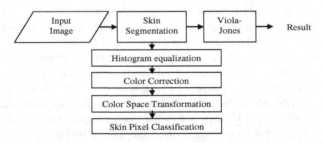

Fig. 1. Workflow of the refined face detection algorithm

The first step in the skin segmentation module performs local histogram equalization. This step is followed by color correction of the image, in order to outbalance its color bias due to the chromaticity of the illumination. We then project the color corrected image from the RGB to YCbCr space and isolate the chromaticity subspace CbCr. After that, we classify the pixels of the image as skin or non-skin and segment the image in skin and background regions. Finally we apply the VJ face detection framework on the selected regions that are likely to contain faces. More details on each step are provided in the rest of the paper.

2.1   The Skin Segmentation Module

Skin segmentation is based on the fact that skin tones form a fairly compact cluster in various color spaces under canonical illumination [16]. Nevertheless, skin segmentation should still be regarded as very challenging due to the intrinsic variability of the skin cluster (ethnicity, individual characteristics), as well as to the extrinsic variability of images (color overlapping with clutter background, capturing device characteristics, illumination variations), which alter the value of the colored pixels.

To start the segmentation process, histogram equalization is applied on local neighborhoods rather than on the whole image, since the latter enhances shadows on the faces, due to high intensity differences in the image.

2.2   Color Correction

Colored illumination biases the chromaticity of objects in a scene towards the color of the light source. This can severely deteriorate the performance of computer vision systems that recognize objects based on their color, like a skin color pixel classifier. In order to circumvent this issue, we use color constancy, i.e. the human visual ability to perceive the color of an object relatively similar under different illuminants. Similarly, computational color constancy aims at maintaining stable colors in an image, regardless of the color of the illuminant. This is a two-phase process and consists of (a) estimating the chromaticity of an illuminant in a scene from the input image, without prior information about the light source, and (b) adapting the image colors, so that they appear as if the image were illuminated with a canonical (neutral) light. The first problem is commonly solved by means of the general expression [21]

$$ I = \frac{1}{k} \left(\iint \left| \nabla^n f_\sigma(x, y) \right|^p dx dy \right)^{\frac{1}{p}}, \tag{1} $$

where $I = (I_R, I_G, I_B)$ is the illuminant color, ∇^n is the n-order derivative, p is the Minkowski norm, and f_σ is the convolution of the image function with a Gaussian filter with scaling parameter σ. Assigning a different triplet of values to the parameters n, p, σ in (1) results in four well-known and widely used color constancy algorithms:

- *Grey World* [4] with $(n,p,\sigma)=(0,1,0)$, which is based on the assumption that the average reflectance in a scene is achromatic.
- *White Patch* [14] or Maximum RGB with $(n,p,\sigma)=(0,\infty,0)$, which is based on the assumption that the maximum reflectance in a scene is achromatic.

- *Shades of Grey* [9] with $(n,p,\sigma)=(0,p,0)$, which is based on the assumption that the pth Minkowski norm in a scene is achromatic.
- *Grey Edge* [21] with $(n,p,\sigma)=(1,p,\sigma)$, which is based on the assumption that the average reflectance of the edges in an image is achromatic.

Once we have estimated the illuminant, we color correct the image using the linear transformation of the von Kries model [22].

In adaptive lighting ambiences, however, there are multiple chromatic light sources of different intensities and in different positions, as mentioned above. This complicates the illuminant estimation. In this study, we benchmark the color constancy algorithms described above in such a multi-color light source environment.

2.3   Color Space Transformation

Color spaces that decouple chromaticity from intensity, such as the YCbCr color space, are considered suitable for color segmentation [6, 19]. Figs. 2 and 3 depict the correlation of Cb and Cr with Y of a skin color cluster, collected from mainly indoor images from the internet, digital cameras and web-cameras. If we look at the figures, we can deduce that the functions Cb(Y) and Cr(Y) remain roughly stable. This implies that both chromaticity components are practically luminance invariant, and as a result, the CbCr space is fairly robust for skin color segmentation. Consequently, we choose to use only the chromaticity components of the YCbCr color space, without the non-linear transformation that Hsu adopted in [10] to remove the effect of very high or very low luminance on the chromaticity components.

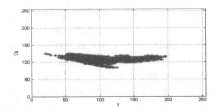   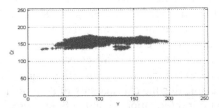

Fig. 2. Skin color cluster in YCb subspace   **Fig. 3.** Skin color cluster in YCr subspace

2.4   Skin Pixel Classification

To detect skin pixels we adopt the naïve Bayesian classifier with histogram technique on the CbCr subspace. A pixel with chromaticity vector $c=(Cb,Cr)$ belongs to the skin class if

$$\frac{p(c\,/\,skin)}{p(c\,/\,nskin)} \geq threshold, \qquad (2)$$

where $p(c/skin)$ and $p(c/nskin)$ are the class-conditional probability distribution functions (pdf) of the skin and non-skin colors, respectively. They are calculated from the occurrence frequency histogram of the skin and non-skin colors. *Threshold* is the value that minimizes the misclassifications to both classes (false positives/false negatives). The threshold theoretically depends on the a priori probabilities of the skin and non-skin classes in the training set; in practice, though, it is determined empirically.

From the output of the skin classifier we create a binary skin mask, assigning 1 to the pixels classified as skin and 0 to the pixels classified as background. Finally, we apply some morphological operators of dilation and erosion to the skin mask in order to include facial features, like the eyes and mouth, and to smooth the binary skin mask. Fig. 4 presents examples of the intensity images after the skin segmentation.

Fig. 4. Intensity images after the skin segmentation module

3   The CI Database of Faces

To verify the robustness of our algorithm under various illumination conditions, we needed a suitable database of faces. From the state-of-the-art face databases only PIE [20] included colored illumination, providing though only the option of blue flash lighting. To be able to test our approach more comprehensively, we produced a new database of faces, the 'CI' (Colored Illumination) database, which suited our particular needs better.

3.1   Characteristics of the Database

The face images were captured in the Experience Lab at Philips High Tech Campus under different colored lighting settings. Sixteen subjects with a variety in skin complexion posed for the database, portraying three different expressions: neutral, frown and smile. This resulted in the CI face database, consisting of 1680 RGB compressed images of faces, which can be divided in two main subgroups: 672 high quality (HQ) images, captured by two identical digital cameras, and 1008 low quality (LQ) images captured by 3 identical web cameras. Table 1 summarizes the main characteristics of each subgroup. Examples of the CI database are presented in Fig. 5.

The use of low-end cameras was motivated from real-world applications, like surveillance or human detection in a room, and offered the opportunity to study the impact of lower-quality images on the detection performance. The high-end cameras captured frontal views of the faces from different horizontal distances. The web cameras were positioned at equal distances, but at different viewing angles (i.e., 0°, 30° and 60° from the frontal view). All cameras were synchronized in order to capture the same situation.

Table 1. Characteristics of the CI database

	Image resolution	Pose	Colored Ambiences	No of images
LQ	1200x1600	frontal side30 side60	neutral cozy 1 cozy 2 activating 1	1008
HQC	2448x 3264	frontal	activating 2 exciting	336
HQF	2448x 3264	frontal	relaxing	336

Table 2. Characteristics of each affective ambience

Ambience	Intensity	Dominant Colors
Neutral	Medium	White
Activating 1	High	Cyan-Blue
Activating 2	High	White-Blue
Cozy 1	Low	Orange-Blue
Cozy 2	Low	Orange-White
Exciting	High	Random colors
Relaxing	Low	Green-Blue

3.2 Lighting Settings

For the experiment we used 7 different lighting settings, each encoded in the remainder of the text with an "affective" name, namely neutral, activating (1&2), cozy (1&2), exciting and relaxing. The affective state of the light settings was recognized by subjects in a separate experiment [13]. These names have no further relevance to the purposes of this study and will only serve as a reference. The different lighting settings can be described based on two main characteristics: their average intensity and dominant colors of the light sources used (see Table 2).

4 Experiments and Analysis

4.1 Experimental Setup

We tested our proposed face detection algorithm on the new CI database and on a subset of the PIE face database[20], consisting of 1165 high resolution images of faces captured in different out-of-plane poses under three different types of illumination: neutral room lighting, only blue flash, and blue flash with neutral indoor lighting. We excluded the few completely dark images of the blue flash subset of the database.

The images were first rescaled, so that the size of all faces was approximately 40 x 40 pixels. We then applied local histogram equalization in neighborhoods of 16 x 16 pixels of the downsized images. Subsequently, we color corrected the images with the four color constancy algorithms discussed above, using the Matlab implementation available online [2]. For the Shades of Grey we used p=5 and for the Grey Edge we set (n,p,σ)=(1,5,2). We quantized the occurrence frequency histograms of skin and non-skin colors in the CbCr subspace using 64 x 64 bins. The threshold used for minimizing the error of the skin pixel classification was calculated theoretically, according to (2), and slightly adjusted upon trial and error on our training image samples.

In the final stage we applied the VJ face detector on the face-candidate areas of the gray scale images. We used the public domain Haar detectors for frontal [15] and profile face detection [3], whose implementation is available in [17] and performance thoroughly tested in [5]. The frontal classifier consisted of 22 stages of weak classifiers, with the minimum size of the face to be detected 24 x 24 pixels, while the profile classifier consisted of 26 stages, with the minimum size of the profiles 20 x 20 pixels. The response of the detector was a square containing the target with some background pixels. The detection was considered correct if the square box included all facial features and its width was not bigger than 4 times the eye distance. Otherwise, the response counted as false positive. We fused the two classifiers for frontal and profile faces in a cascade structure, only using the profile classifier if the frontal one failed, for the main reason that the former has been proven more accurate than the latter [5]. We separated our test set of images into eight subgroups (four subcategories per database), as illustrated in detail in Figures 6a-6b.

4.2 Performance of the Color Constancy Algorithms

The performance of the algorithms is compared in terms of *accuracy*, defined as

$$accuracy = \frac{DR}{1+FP}, \tag{3}$$

where DR is the detection rate and FP the false positives rate. Table 3 summarizes the accuracy of the four color constancy algorithms, tested on the images with frontal faces of the LQ dataset, indicating the added value of color correction. Concerning the final face detection accuracy, the Shades of Grey color constancy algorithm achieves the highest average result and performs well in all different colored ambiences. The simplest Grey World performs equally well in many cases in terms of detection rate, but it has a higher number of false positives. The White Patch outperforms in ambiences with high intensity, but is unsuitable at medium and lower intensities. Hence, for the rest of the tests we applied skin segmentation after color correction with Shades of Grey.

4.3 Comparison of the Refined Algorithm with Other Face Detection Methods

Figure 6 evaluates the performance of the proposed method (Shades of Grey-based skin segmentation previous to the VJ detector) under challenging image quality, pose, and illumination conditions. For comparison, we also report the accuracies of other state of the art face detectors, namely the basic VJ algorithm [23] and Erdem et al.'s method [8], which refines the VJ results with subsequent Grey World-based skin segmentation. The proposed method seems to be more sensitive to pose rather than quality or illumination. The reason for this is that we have used a classifier mainly trained to detect frontal or nearly frontal faces. Strong out-of-plane-rotations that lead to occlusions of facial parts, like one eye, may cause the classifier to fail. Even so, the skin segmentation improves the detection performance with respect to [8, 23]. The second most significant factor is the low quality of the web-cameras, as expected. This is partly due to the false positives, resulting from the cluttered background, captured from the larger field of view of the web-cameras. Another reason is that the low-end cameras do not operate color adaptation mechanisms, as opposed to the high-end ones. Challenging multicolor illumination can decrease the accuracy of the VJ detector more than 20%. In these cases, the skin segmentation module can improve the accuracy significantly, up to 10-15%.

Fig. 5. Examples of the different illuminated ambiences of the CI database; from left to right: a) cozy 1, b) activating 2, c) activating 1(a-c: LQ), d) exciting (HQ F), and e) cozy2 (HQC)

Finally, it should be stressed that the moment at which the skin segmentation filter is applied is crucial (Fig. 6a): applying VJ on detected skin regions yields better performance than using a post-VJ skin segmentation filter [8] to refine the results   (in terms of FP).

Table 3. Final accuracy of the refined face detection algorithm for different color constancy algorithms. Tests on 336 LQ images with frontal faces of size 40x40

	GW	WP	SG	GE	No CC	VJ
neutral	0,75	0,82	0,78	0,72	0,77	0,53
cosy1	0,51	0,35	0,56	0,35	0,47	0,47
cosy2	0,80	0,71	0,80	0,72	0,74	0,69
act1	0,94	1	0,96	1	0,92	0,77
act2	0,87	0,87	0,89	0,89	0,87	0,66
exc	0,69	0,69	0,69	0,70	0,67	0,61
relax	0,43	0	0,63	0,10	0,02	0,46
average	**0,71**	**0,63**	**0,76**	**0,64**	**0,64**	**0,60**

GW: Grey World, **WP:** White Patch, **SG:** Shades of Grey, **GE:** Grey Edge, **No CC:** no color correction, **VJ:** Viola-Jones without the skin segmentation module

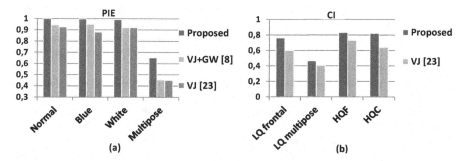

Fig. 6. Accuracy of the VJ algorithm, VJ after skin segmentation module with SG (proposed method) and VJ before skin segmentation with GW [8]: tests on the PIE and CI (comparison only of the first two methods) databases: (a) from the Pie: 867 with frontal faces under indoor illumination, flash and indoor illumination and only blue flash, and 298 with multi-pose faces (b) from the CI: 336 LQ with frontal faces, 672 LQ with multi-pose faces, 672 HQ with frontal faces from the closest and furthest cameras. Faces of size 40x40

5   Conclusions

We have presented a face detection algorithm that retains performance under colored illumination, using a skin segmentation module before the VJ algorithm. The major contribution of the skin mask is the refinement of the detection by eliminating false positives. A pure appearance based system cannot cope with the false positives very efficiently, because it only uses information from the grey-scale image.

This study can be generalized to every appearance-based face detection method in ambiences with colored illumination. More effort can be put on refining the skin mask, taking into account pair-wise dependencies between adjacent pixels. In addition, shape constraints for the face can exclude non-face skin areas (e.g. hands) or skin-like regions (e.g. wooden furniture).

Finally, the parameters of the color constancy algorithms can be adjusted automatically to the different colors of illumination, exploiting prior information and intrinsic properties of the image. Even so, this study highlights the importance of color correction even with an overall optimal color constancy algorithm, before skin segmentation, and quantifies the added value of the latter to the accuracy of an appearance based face detector.

References

1. Aarts, E.: Ambient Intelligence – "Visualizing the future". In: Proc. Conf. Smart Objects & Amb. Intelligence (2005)
2. Gijsenij, A., Gevers, T.: http://colorconstancy.com
3. Bradley, D.: Profile face detection (November 5, 2003),
   http://www.davidbradley.info/publications/
   bradley-iurac-03.swf

4. Buchsbaum, G.: A spatial processor model for object color perception. J. of the Franklin Institute 310(1), 1–26 (1980)
5. Castrillón, M., Déniz, O., Hernández, D., Lorenzo, J.: A comparison of face and facial feature detectors based on the Viola-Jones general Object detection Framework. Machine Vision and Applications 22, 481–494 (2011)
6. Chai, D., Ngan, K.N.: Face segmentation using skin color map in videophone applications. IEEE Trans. Circuits and Systems for Video Technology 9(4), 551–564 (1999)
7. Chen, H.Y., Huang, C.L., Fu, C.M.: Hybrid-boost learning for multi-pose face detection and facial expression recognition. Pattern Recognition 41(3), 1173–1185 (2008)
8. Erdem, E., Ulukaya, S., Karaali, A.: Combining Haar feature and skin color based classifiers for face detection. In: IEEE Int'l Conf. on Acoustics, Speech and Signal Processing (ICASSP), pp. 1497–1500 (2011)
9. Finlayson, G., Trezzi, E.: Shades of gray and color constancy. In: Color Image Conference, pp. 37–41 (2004)
10. Hsu, R.L., Abdel-Mottaleb, M., et al.: Face detection in color images. IEEE transactions on pattern analysis and machine intelligence 24(5) (2002)
11. Intel, Intel Open Source Computer Vision Library, http://sourceforge.net/projects/opencvlibrary/
12. Kim, B., Ban, S.-W., Lee, M.: Improving adaBoost based face detection using face-color preferable selective attention. In: Fyfe, C., Kim, D., Lee, S.-Y., Yin, H. (eds.) IDEAL 2008. LNCS, vol. 5326, pp. 88–95. Springer, Heidelberg (2008)
13. Kuijsters, A., Redi, J.A., de Ruiter, B., Heynderickx, I.: Effects of ageing on atmosphere perception. In: The Process of Being Published in Proceedings of Experiencing Light (2012)
14. Land: The retinex theory of color vision. Scientific American 237(6), 108–128 (1977)
15. Lienhart, R., Maydt, J.: An Extended Set of Haar-like Features for Rapid Objects Detection. In: IEEE ICIP 2002, vol. 1, pp. 900–903 (2002)
16. Phung, S.L., Bouzerdoum, A., Chai, D.: Skin Segmentation using color pixel classification: Analysis and Comparison. IEEE Transactions on Pattern Analysis and Machine Intelligence 27(1) (2005)
17. Reimondo, A.: Haar cascades repository (2007), http://alereimondo.no-ip.org/OpenCV/34
18. Riemersma-van der Lek, R.F., Swaab, D.F., et al.: Effect of Bright Light and Melatonin on Cognitive and Noncognitive Function in Elderly Residents of Group Care Facilities: A Randomized Controlled Trial. JAMA 299(22), 2642–2655 (2008)
19. Sobottka, K., Pitas, I.: A Novel Method for Automatic Face Segmentation: Facial Feature extraction and tracking. Signal Processing: Image Comm. 12(3), 263–281 (1998)
20. Sim, T., Baker, S., Bsat, M.: The CMU Pose, Illumination, and Expression (PIE) Database. In: Proc. Int'l Conf. Face and Gesture Recognition, pp. 46–51 (2002)
21. de Weijer, J.V., Gevens, T., Gijsenij, A.: Edge based color constancy. IEEE Transactions on Image Processing 16(9), 2207–2214 (2007)
22. von Kries, J.: Chromatic adaptation. Festschrift der Albrecht-Ludwigs-Universität (1902); Translation: D.L. Mac Adam, Colorimetry-Fundamentals, SPIE Milestone Series, vol. MS 77, (1993)
23. Viola, P., Jones, M.: Rapid object detection using a boosted cascade of simple features. In: Proc. of the Conf. on Computer Vision and Pattern Recognition, CVPR 2001 (2001)
24. Zhang, C., Zhang, Z.: A survey of recent advances in face detection. Microsoft Research, 202.114.89.42, 5 (2010)

Easy Carbon Offsetting
and Trading with Ambient Technologies

Ichiro Satoh

National Institute of Informatics
2-1-2 Hitotsubashi, Chiyoda-ku, Tokyo 101-8430, Japan

Abstract. This paper presents a novel approach to carbon credit trading with ambient computing technologies, particularly RFID or barcode technology. It introduces RFID tags as certificates for the rights to claim carbon credits in carbon offsetting and trading. It enables buyers, including end-consumers, that buy products with carbon credits to hold and claim these credits unlike existing carbon offsetting schemes. It also supports the simple intuitive trading of carbon credits by trading RFID tags coupled to the credits. The approach has been already constructed and evaluated with real customers and real carbon credits in a real supply chain.

1   Introduction

Ambient intelligence is useful to build a sustainable world. This paper introduces RFID tags or barcodes, which are ambient intelligence technologies, to reduce greenhouse gases (GHGs), including carbon dioxide (CO_2). *Carbon credits* provide an economical approach to reducing the amount of GHG emissions, where carbon credits are generated by the reduction of CO_2 emissions in sponsoring projects, which increase CO_2 absorption, such as renewable-energy, energy-efficiency, and reforestation projects. Although carbon credits themselves do not reduce the amount of CO_2 emissions around the world, they are important incentives for GHG reduction projects. Many companies have also sold products with the amount of carbon credits equivalent to the amount of GHGs emitted due to the use or disposal of products so that the credits have been used to offset GHGs emissions. There are a variety of products on the market with carbon credits, e.g., automobiles, disposable diapers, and toys. For example, from September 2007, Lufthansa began offering its customers the opportunity of offsetting carbon emissions through voluntarily donating carbon credits to mitigate the amount of CO_2 emitted due to the actual average fuel consumption per passenger.

However, carbon offsetting poses several serious problems that result from carbon credit trading. Carbon credits are usually acquired through *carbon credit trading* between countries or companies, or in markets via professional traders, called carbon traders or agencies. However, existing trading schemes are too complicated for non-professional traders, individuals, or small and medium-sized enterprises, to participate in. Furthermore, the minimal unit of existing credit trading is usually more than one hundred or one thousand tonnes of CO_2 emissions, whereas the amount of GHGs emitted due to the use or disposal of consumables is less than one kilogram.

A. van Berlo et al. (Eds.): *Ambient Intelligence – Software & Applications*, AISC 219, pp. 205–212.
DOI: 10.1007/978-3-319-00566-9_26   © Springer International Publishing Switzerland 2013

This paper aims at enabling a small amount of carbon credits attached to products to be transferred to endconsumers who buy these products and carbon credits to be easily traded. The key idea behind our proposed approach is to use RFID tags (and barcodes) as certificates for the rights to claim carbon credits and middleware for ambient intelligence [7]. We designed an architecture for managing RFID-enabled carbon credit offsetting and trading. The architecture was constructed and evaluated with real carbon credits in a real supply-chain system.

2   Related Work

Several researchers have explored computing technology to make a contribution to the environment. For example, Persuasive Appliances [6] was an interface system to provide feedback on energy consumption to users. PowerAgent [1] was a game running on mobile phones to influence everyday activities and minimize the use of electricity in the domestic settings. UbiGreen [3] was an interactive system running on mobile phones and gave users feedback about sensed and self-reported transportation behaviors to reduce CO_2 emissions from the transportation sector.

There have been several projects that have used sensing systems to manage warehouses and logistics to reduce CO_2 emissions. Ilic et al. [4] proposed a system for controlling the temperature of perishable goods to reduce GHG emissions. Dada, et al. [2] proposed a system for accurately quantifying GHG emissions to calculate carbon footprints and communicate the results to consumers through sensing systems. The system also planned to use EPCglobal RFID tags to trace carbon footprint emissions at higher stages of the supply chain.

3   Basic Approach

This paper proposes an approach for enabling carbon credits attached to products to be transferred to consumers who buy these products. Our approach introduces RFID tags (or barcodes) as carbon credits for the rights to claim credits in carbon offsetting, because RFID tags (or barcodes) are used in supply chains. In fact, our approach can use the RFID tags (or barcodes) that have already been attached to products for supply chain management. The approach was designed as a complement to existing supply management systems. It therefore has nothing to do with the commerce of products themselves. It also leave the transfer of carbon credit between companies with existing carbon trading systems, because commerce for carbon credits must be processed by certificated organizations. Instead, the approach is responsible for attaching carbon credits to RFID tags and claims for carbon credits. The approach should support emission credits and caps in a unified manner. It also should not distinguish between products for end-consumer and others, because non-end-consumers may buy products for end-consumers. Some readers may think that our approach is trivial. However, simplicity and clarity are essential to prompt most people and organizations to participate or commit to activities to reduce GHG emission by carbon offsetting.

Our approach satisfies the following main requirements: 1) The approach needs to encourage industries and homes to reduce GHG emissions. It also needs to be compliant with regulations on carbon offsetting. 2) Simplicity must be a key concern in minimizing operation costs, because it tends to be in inverse proportion to cost. This is needed for people and organizations to understand what is required of them. 3) Any commerce scheme provides the potential to advantage some participants at the expense of others. The approach enables organizations or people that reduce more GHG emissions to be rewarded with greater advantages. 4) The values of carbon credits, particular emission credits tend be varied. The amounts, expiration dates, and sources of all carbon credits, which may be attached to products, need to be accessible. 5) When consumers purchase products with carbon credits, they should easily be able to own the credits without any complicated operations to authenticate them. 6) Product commerce in the real world is often done in warehouses and stores, where networks and electronic devices may not be available. Our approach itself should be available offline as much as possible.

4 Design

Our approach introduces RFID tags (or barcodes) as carbon credits for the rights of emitters to claim credits in carbon offsetting and trading, because RFID tags (or barcodes) are used in supply chains. In fact, our approach can use the RFID tags that have already been attached to products to manage supply chains. The approach was designed to complement existing supply management systems. It therefore has nothing to do with the trading of the products themselves. It also leaves the transfer of carbon credits between companies to existing carbon trading systems, because carbon trading must be processed by certificated organizations. Instead, the approach is responsible for attaching carbon credits to RFID tags and claims for carbon credits. The approach should support emission credits and caps in a single manner. It should also not distinguish between products for endconsumers and others, because non-endconsumers may buy products for endconsumers.

4.1 RFID Tags as Certificates to Claim Carbon Credits

One of the most novel and significant ideas behind our approach is to use RFID tags themselves, rather than their identifiers, as certificates for carbon credits. This is because it is difficult to replicate or counterfeit RFID tags whose identifiers are the same, because their identifiers are unique and embedded into them on the level of semiconductors. That is, we can assume that one identifier will always be held in at most one RFID tag.

To claim carbon credits dominated by RFID tags, we need to return these RFID tags to the stakeholders that assigned carbon credits to the tags. This is because there is at most one RFID tag whose identifier is the same. RFID tags can be used as certificates for carbon credits. For example, when sellers want to attach carbon offsetting credits to products, they place RFID tags on them that represent the credits for the products. Our approach couples carbon credits with RFID tags themselves, instead of the identifiers of the RFID tags. Therefore, purchasers, who buy the products, tear the RFID tags from them and return the tags to the sellers (or the stakeholders of the credits). When the sellers receive the RFID tags from the purchasers, they transfer the credits to any accounts for payments that the purchasers specify.

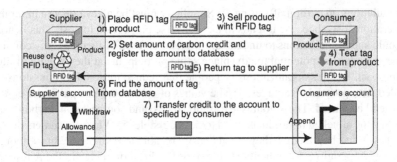

Fig. 1. RFID-based attachment of carbon credits to products

Figure 1 explains our approach to attach carbon credits to products with RFID tags, which involves seven steps

1) A seller places an RFID tag on a product (or a volume of products) if the product has no tag.
2) It sets a certain amount of carbon credits for offsets for a product and registers the amount and the identifier of the tag in a database.
3) It sells the product with the RFID tag to a purchaser.
4) The purchaser tears the tag from the product that it has bought.
5) It only returns the tag with information about the account that the credit should be paid to, to the seller.
6) The seller receives the tag and then finds the amount of carbon credits coupled to the tag in the database.
7) It transfers the amount to the account specified by the purchaser and removes information on the identifier from the database so that the tag can be reused.

4.2   Carbon Credit Trading with RFID Tags

When a purchaser has torn an RFID tag from a product, which might have been attached to a product that he/she purchased, our approach permits the purchaser to resell the tag to others (Figure 2). Instead, the new holder of the tag can claim the carbon credits attached to the tag from the stakeholder of these credits or resell them to someone else. Note that trading RFID tags corresponds to trading carbon credits.

To offset GHG emissions according to the Kyoto protocol, we must donate certified carbon credits to the government via a complicated electronic commerce system. Our approach provides two approaches to carbon offsetting. The first is to simply donate RFID tags coupled to certified carbon credits to the government. For example, people can simply throw RFID (unsigned) tags into mailboxes to contribute to reducing GHG emissions in their home countries. The government then gathers the posted tags. The second is to explicitly specify the certificated cancellation account of the government as the account that the credit should be paid into.

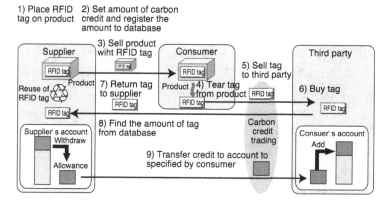

Fig. 2. RFID-enabled trading of carbon credit

4.3 Carbon Credit Agent

Anyone can access information about the credits attached to the products, because the credits are transferred to purchasers who return the tags themselves to the sellers. The sellers should provide information about the credits, e.g., their amounts, expiration dates, and sources. This approach provides web-based agents, called *carbon credit agents*, to enable customers to access the information. The agents are running as programmable entities on web servers specified at URLs in RFID tags or barcodes and can be extended with the ability of artifical or ambient intelligence. When customers can read RFID tags or barcodes with web-enabled terminals, they see information on the credits attached to the tags or barcodes.

5 System

Our approach assumes that sellers at steps in a supply chain will sell their products to customers, including raw materials and components, with RFID tags coupled to carbon credits.

- Our approach requires each RFID tag to have its own unique read-only identifier. Most RFID tags used in supply chain management already have such identifiers.
- To support carbon offsets, the amount of credits attached to a product need to be equivalent to the total or partial amount of CO_2 emissions resulting from the use or disposal of the products.[1]

The system in Figure 3 is self-contained but it may cascade from upstream to downstream along a supply chain. Some readers may worry that returning RFID tags to their stakeholders is more costly than returning the identifiers of tags via a network. There

---

[1] This approach itself is intended to leave the amount of credits attached to a product at the stakeholder's discretion, because the credits can be an incentive to sell the product.

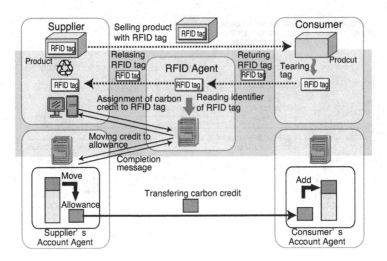

Fig. 3. System architecture

are two flows that are opposite to each other between sellers and purchasers at each stage in real supply chains; the flows of products and the flows of receipts or containers for the products. Our approach can directly use the latter flow to return tags from purchasers to sellers. Therefore, our cost and extra CO_2 emissions are small. Actually, returnable containers, which deliver parts or components from sellers and then return them to sellers, are widely used in real supply chains.

- Each seller has at least one carbon credit account entrusted to agents for carbon credit accounts. It has RFID tag reader systems to read the identifiers of RFID tags. If a seller consigns one or more RFID agents to manage RFID tags for carbon credits, they need a database to maintain which RFID agent will manage each of the RFID tags.
- Each purchaser may have at least one carbon credit account entrusted to agents for carbon credit accounts. It buys products that RFID tags have attached to them for carbon credits from sellers or traders. It needs RFID tag reader systems, when it intends to access information about carbon credits.
- Agents for carbon credit accounts, simply called *account agents*, may be existing certified carbon providers. They have two databases. The first maintains carbon credit accounts and the second maintains information about assigned credits. They can only be connected to certain RFID agents and other account agents through authenticated and encrypted communications.
- An RFID agent has a database to couple the identifiers of RFID tags and information about carbon credits. The agent may lease RFID tags, which may already have been assigned a certain amount of credits to sellers.

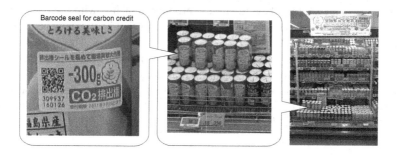

Fig. 4. Beverage with Barcode for carbon credits

6   Early Experience

The experiment was an early case study on the proposed approach, but was carried out on a supply chain for beverages it was evaluated at several steps in the supply chain, including beverage companies (e.g., Pokka and Fujiya), a supermarket (Kitasuna branch of Ito-yokadou) [2], and a carbon credit agency (Mitsubishi UFJ Lease). It was carried out for two weeks from 9 am to 10 pm and more than five thousand goods were sold with carbon offset credits in this experiment. The supermarket opened the returnable containers or cardboard boxes containing the cans. It attached a barcode seal on the cans and sold them to endconsumers, where each barcode seal displayed small amount of J-VER carbon credits, because the price of each RFID tag was relatively more expensive than the price of a can. Each barcode was formatted in a 2D barcode, called QR code, and consisted of its own identifier, the weights of carbon emission credits assigned to it, and the address of the management server. Figure 4 shows beverage cans with barcodes in a showcase at the supermarket.

Endconsumers bought cans and collected barcode seals as their carbon credits. We supported two cases to reclaim credits in the proposed approach.

- The first was for endconsumers to return barcode seals to the supermarket to reclaim credits. Cashiers could distinguish between original or imitation seals, because they received the seals themselves. Therefore, even when someone read the barcodes attached to the cans, the endconsumers who bought the cans could reclaim the credits.
- The second was for endconsumers to read barcodes by using their own scanners, e.g., cellular phones with cameras, and they then sent the information to the server specified in the barcode. As some might peel off the barcode attached on the cans in stores and illegally reclaim carbon credits. The barcodes were concealed by covering them with other seals.

We used the former in our approach. Few endconsumers participated in the latter, because most endconsumers wanted to immediately reclaim their credits. The former also enabled consumers to access information about carbon credits by reading the barcodes

---

[2] The supermarket is one of the biggest in Tokyo area.

with cellular phones before they bought the products attached with the barcodes. The experiment assumed that retailers bought barcode seals that had already been assigned to small amounts of carbon emission credits, like postage stamps. This is because small retailers might not have any terminals. The sales volumes of cans with carbon credits in two weeks was three times more than usual at the supermarket. Thirty-five percent of barcodes were returned to the supermarket by customers who claimed the credits. The experiment enabled consumers to offset their CO_2 emissions by using the carbon credits they reclaimed from the barcodes.

7   Conclusion

The approach proposed in this paper can be proposed to solve serious problems with carbon credits, offsetting, and trading. The key idea underlying our approach is to introduce RFID tags (or barcodes) as physical certificates for the rights to claim carbon credits, including carbon emission credits and caps. When purchasers buy products with credits for carbon offsets, they can claim the credits by returning the RFID tags (or barcodes) coupled with the credits to stakeholders, e.g., sellers or agencies, without the need for any complicated authentication. The approach can treat carbon credit trading as the trading of RFID tags. The approach was constructed to complement existing systems of supply chain management and existing systems of carbon credit trading (or barcodes).

References

1. Bang, M., Gustafsson, A., Katzeff, C.: Promoting new patterns in household energy consumption with pervasive learning games. In: Proceedings of 2nd International Conference on Persuasive Technology, pp. 55–63 (2007)
2. Dada, A., Staake, T., Fleisch, E.: The Potential of UbiComp Technologies to Determine the Carbon Footprints of Products. In: Proceedings of Pervasive Computing 2008 Workshop on Pervasive Persuasive Technology and Environmental Sustainability, pp. 50–53 (2008)
3. Froehlich, J., Dillahunt, T., Klasnja, P., Mankoff, J., Consolvo, S., Harrison, B., Landay, J.A.: UbiGreen: investigating a mobile tool for tracking and supporting green transportation habits. In: Proceedings of the 27th International Conference on Human Factors in Computing Systems (CHI 2009), pp. 1043–1052. ACM (2009)
4. Ilic, A., Staake, T., Fleisch, E.: Using Sensor Information to Reduce the Carbon Footprint of Perishable Goods. IEEE Pervasive Computing 8(1), 22–29 (2009)
5. IPCC fourth assessment report: Climate change 2007. Technical report, IPCC (2007), http://www.ipcc.ch/
6. McCalley, T., Kaiser, F., Midden, C., Keser, M., Teunissen, M.: Persuasive Appliances: Goal Priming and Behavioral Response to Product-Integrated Energy Feedback. In: IJsselsteijn, W.A., de Kort, Y.A.W., Midden, C., Eggen, B., van den Hoven, E. (eds.) PERSUASIVE 2006. LNCS, vol. 3962, pp. 45–49. Springer, Heidelberg (2006)
7. Satoh, I.: An agent-based framework for context-aware digital signage. In: Augusto, J.C., Corchado, J.M., Novais, P., Analide, C. (eds.) ISAmI 2010. AISC, vol. 72, pp. 105–112. Springer, Heidelberg (2010)
8. Stern, N.: The Economics of Climate Change: The Stern Review. Cambridge University Press (2007)

Ethically Intelligent? A Framework
for Exploring Human Resource Management Challenges
of Intelligent Working Environments

Céline Ehrwein Nihan[1] and Katharina Kinder-Kurlanda[2]

[1] University of Applied Sciences in Business and Engineering Vaud (HEIG-VD),
Av. des Sports 20, CH-1401 Yverdon-les-Bains, Switzerland
celine.ehrwein@heig-vd.ch
[2] GESIS Leibniz-Institut für Sozialwissenschaften,
Unter Sachsenhausen 6-8, 50667 Köln, Germany
katharina.kinder-kurlanda@gesis.org

Abstract. With advances in the development of intelligent environments (IEs) social scientists and ethicists have begun to gauge the social impact of the diverse emerging technology assemblages and to work with developers and stakeholders in order to improve design and deployment of such technologies. Research conducted to better understand specific socio-technical settings faces multiple challenges due to the complexity of most scenarios, making it imperative to approach the attempt of understanding IEs both from multi-disciplinary perspectives and to include practitioners and managers of IEs. In this paper we introduce our framework for setting up a competence center (CC) for defining and managing socio-ethical challenges of intelligent *working* environments (IWEs). We detail practical steps to successfully build a CC in order to allow other research projects to adopt some of these steps or the whole framework to positively anticipate and manage socio-ethical challenges of IWEs.

1 Introduction and Problem Setting

Currently numerous research projects are being conducted in the field of Information and Communication Technology (ICT) in order to develop environments which can record, analyse and handle ambient data from diverse sources and which are intended to meet the needs of their users automatically, in due time, and in a personalized and intelligent manner [3, 12]. These efforts are expected to radically transform our perception of reality and the organization of our everyday life [7], including the organization of work. In particular, intelligent environments (IEs) may be used by companies for managerial purposes providing accurate, real-time information. IEs can be integrated into workplaces to measure and improve employees' working conditions, well-being and performance, and, last but not least, to optimize the organizations' productivity.

Studies related to new intelligent working environments (IWEs) have focused on technical difficulties [14], or on the implications of ubiquitous computing for knowledge management [11]. Scholars have sketched scenarios of the future

A. van Berlo et al. (Eds.): *Ambient Intelligence – Software & Applications*, AISC 219, pp. 213–219.
DOI: 10.1007/978-3-319-00566-9_27   © Springer International Publishing Switzerland 2013

workplace [2], have conducted studies on the acceptance of IWEs [13], and have highlighted possible organizational issues [1]. The specific question of the ethical and human resource (HR) management issues raised by IEs in the workplace has not yet been explored. This gap is all the more striking since the first applications related to the workplace are arriving on the market[1]. Yet, neither employees nor (HR) managers seem to be sufficiently aware of the radical changes that are occurring and of their own role in the implementation of these developments. Furthermore, the socio-political impact of these new technologies and questions of whether and how to adapt legal frameworks to the development of IWEs are starting to be assessed. An interdisciplinary examination of these issues with scholars and professionals is urgent if we want to anticipate, support and frame the technological developments to come.

2   General Objectives and Expected Outcomes

With the perspective outlined in the previous section in mind, we have started a project with the following aims:

- The exploration of the socio-ethical and HR issues raised by the arrival of IEs in the workplace;
- The raising of awareness in (HR) managers and designers of work environments with respect to IWE challenges, and;
- The elaboration of tools intended to help (HR) managers to take part in the developments in an active and constructive manner.

In the short term we expect to develop recommendations to (HR) managers and environment designers with regards to developing strategies and measures for organizations to meet the ethical and HR challenges of IWEs. We also seek the integration of findings into graduate and postgraduate teaching programs in order to raise awareness in engineering and management students of the challenges they will face. *In the long term* we aim to develop further interdisciplinary projects related to the design of IWEs and to support the elaboration of recommendations for political authorities.

3   Methodology

Defining challenges of IWEs requires thinking about how such a definition process can be accomplished. In the following section we first show the methodological challenges we faced in our project, then present building a competence center (CC) as a solution to these challenges, and finally explain the main steps completed since the project was initiated at the end of 2011[2].

---

[1] See for example the 3D job interviews simulator developed by the Centre de réalité virtuelle de Clermont-Ferrand and the consulting company Athalia (www.aprv.eu) or the intelligent fireman hood developed by the firm Bodysens (www.bodysens.com).

[2] The project was initiated by Céline Ehrwein Nihan from of the Human Resources and Management Unit of the University of Applied Sciences in Business and Engineering Vaud (HEIG-VD) and Bernard Baertschi of the Institute for Biomedical Ethics of the University of Geneva.

3.1 Requirements and Challenges

Deployment of an Interdisciplinary Perspective

Often the decision to implement IWEs in companies is in the hands of top management and HR departments. However, IWEs do not solely concern (HR) managers. The rise of ubiquitous computing in the workplace implies a *wide variety of stakeholders* (certainly managers, but also workers, engineers, designers, etc.), who make sense of technologies from various viewpoints and who pursue different – and sometimes conflicting – interests [8, 10]. Furthermore, many ethical and legal issues, such as power balance, patent law, the right to privacy, accountability, etc., which come into play with IWEs [1, 5, 9], put at stake *socio-ethical regulations and axiological frameworks that go far beyond companies*. Finally, the development of IWEs involves *fundamental cultural as well as anthropological changes* [4, 7]. In other words, the settings of IWEs are complex: they raise managerial, epistemological, socio-technical and socio-ethical challenges. Handling the complexity of these challenges should neither be bound by disciplinary concerns nor be assumed solely on the level of individual actors' personal ethical obligations. This *implies an interdisciplinary perspective and a shared responsibility*. Our intention is to make such an interdisciplinary exchange possible and to support (HR) managers and environment designers to face future challenges.

Development of Cross-Institutional Collaborations

One of the risks related to research of future IWE developments lies in the emergence of a gap between scholarly reflections and the needs and expectations of companies and professionals [10]. This gap may compromise the relevance of research initiatives. Therefore it is essential to ensure a dissemination of research results in academic communities as well as amongst professionals in order to *facilitate comprehension and implementation* of our work. Bridging the gap between researchers and practitioners is all the more important as many IWE applications are only now emerging. Both practitioners' and developers' perspectives are required to anticipate new challenges. Thus, to prevent *the risk of being cut off from professional realities and practices* we maintain a strong connection with companies and (HR) managers.

Setting of a Common Research Structure

Interdisciplinary work and cross-institutional collaborations are daring undertakings. There is a risk of profound epistemological misunderstandings amongst scholars with different backgrounds as well as of misunderstandings between academics and practitioners. Therefore the *setting of a common structure for the research* is essential. *Institutionalised long term relations* among both scholars and professionals need to be established. It is also necessary to develop a *shared methodology* to guarantee and improve the focus of the research group in the long term. Facing these challenges we decided to work, towards building a CC, based in Switzerland but with an international outlook.

3.2 Building a Competence Center

The CC includes researchers and practitioners from various fields and disciplines and is intended to enhance the quality, visibility and accessibility of the project and its results. In the following section we outline our framework for building a CC by detailing the individual steps taken.

Steps Taken to Ensure an Interdisciplinary Perspective

One of the first steps was to gather scholars from different academic fields who were experienced and interested in intelligent technologies and who represented a variety of renowned research centers and university departments[3]. We recruited specialists who between them covered a wide range of fields, among them anthropology of technology; biotechnology and nanotechnology; corporate and HR management; bioethics, social ethics and ethics of politics; patents and new technology law; sociology of work; and science and technology studies. We then established a scientific and organisational committee at the core of the CC to be in charge of the direction and implementation of the project. The composition of the committee met the demands of interdisciplinarity.

Steps Taken to Ensure Cross-Institutional Collaborations

In June 2012, the Human Resources and Management Unit of the HEIG-VD, which first launched the project, was invited by the Association for Human Resources Management (HR-Vaud) for a conference on the theme of smart technologies[4]. This meeting allowed for an assessment of the relevance of the topics for professionals [6]. A partnership was concluded between HR-Vaud and the project committee. Two HR managers actively participate in the research project and the association supports the dissemination of project results among professionals.

With the aim of an expansion of the project to a broader societal horizon two centers for technology assessment (TA-Swiss in Switzerland and Rathenau Institute in the Netherlands) specialized in public participation methods and in the elaboration of recommendations for political authorities were invited to join the research group.

At the present time discussions for an implementation test in a company are ongoing, which involve the project committee, the HR department of a multinational company and two research centres developing IE applications. We expect these discussions to offer a common case study application, also intended to focus the research.

Steps Taken to Ensure the Setting of a Common Research Structure

The time frame of the project was set as a three year period at the end of which the CC will be fully operational. To ensure the collaboration of people with different academic backgrounds and perspectives through a focused structure we combined three different methods of collaboration:

---

[3] For example, Institute for Social Sciences of the University of Lausanne, Institute for Information and Communication Technologies of the HEIG-VD, etc.

[4] A short version of this conference contribution was also presented in October 2012 at the Swiss Exposition for Human Resource Management: www.salon-rh.ch

- Interdisciplinary workshops (1-2 days every 5 months) to establish the framework and the main lines of the research;
- Individual work to allow each scholar to deepen specific problems and to conduct field investigations;
- Discussions with (HR) managers to guarantee the integration of practitioners' needs and expectations throughout the process.

The workshops were carefully prepared (establishment of a program, clear cut objectives and instructions) by the committee members and participants were requested to submit work in advance as a basis for the discussion. One or two participants would also be invited to present a paper related to their individual research in order to gain momentum for the discussion. Each workshop resulted in a report. During the workshops the committee members would pay particular attention to the interdisciplinarity of the discussion (e.g. speaking slots), the achievements of the pre-defined objectives and the integration of the points of agreement/disagreement in the workshop reports.

In addition, an on-line platform was created to allow the project members to share information required for the research. This platform is being used as support for a project database.

4   First General Outcomes and Steps to Take in the Future

4.1 First General Outcomes

The research group met twice during the first half of 2012 to assess the relevance of the research topic, to define general objectives and to determine the common methodology. In addition to the points mentioned above (see 3.2), the first main outcomes are:

- Elaboration of a common definition of IWEs[5];
- Clarification of the degree of involvement in the project realisation for each group member;
- Establishment of a list of research interests studied in the project;
- Establishment of criteria for the selection of applications to be studied in interdisciplinary research;
- Selection of two applications representative of IWEs and their related ethical and managerial issues[6].

With regards to the success of the measures taken to ensure an interdisciplinary perspective and cross-institutional collaboration we found the workshops with their specific aim of allowing all present viewpoints to be heard to be the most effective

---

[5] The definition which has been elaborated is: "Working environments fitted with ubiquitous computing system(s), often imperceptibly, which records, integrates, correlates and analyzes ambient data from diverse sources and is intended to meet the needs of the stakeholders automatically, in due time and in a personalized and intelligent manner."

[6] The selected applications relate to health care management and to clothing with sensors.

tool in facilitating a collaboration between people who may not normally have interacted with each other. Translating the workshop discussions into results became possible through the mix of individual, small-group and large-group activities. However, we find room for improvement in this translation process: In hindsight we think we may have underestimated the necessity of keeping questions more closed and result-oriented rather than open and explorative. The latter, while necessary in early phases of the project, can later on lead to questions being raised repeatedly and to continued discussions of the *mode* of discussion. Therefore, we are aiming to keep an even closer eye on carefully defining the intended results of every activity to ensure measureable progress with the project.

4.2 Main Steps to Take in the Future

While the project is still in its first stages, we intend to pursue it through further organization of workshops, individual work and discussions between scholars and (HR) managers. These are expected to result in an interdisciplinary definition and analysis of a) global socio-ethical issues of IWEs, b) ethical issues for (HR) managers and IWE designers in particular and, c) HR management issues, all indispensable prerequisites to the development of relevant recommendations to (HR) managers and IWEs designers. Having developed these recommendations, the CC will be in a position to establish a roadmap for the development of further projects. For example, we plan the organization of a civic forum on IWEs in order to support the elaboration of recommendations to political authorities.

5   Conclusion

In this paper we have shown how we are building a CC to engender interdisciplinarity, cross-institutional collaboration and a common research structure in order to facilitate timely and relevant research of the socio-ethical issues that (HR) managers and designers face with regards to IWEs. We have introduced the framework that facilitates the research. Our aim in this is to allow other projects to implement our framework, or parts of it, in order to include an exploration of socio-ethical issues in their own projects.

References

1. Boos, D., Guenter, H., Grote, G., Kinder, K.: Controllable Accountabilities The internet of things and its challenges for organisations. Behav. & Inf. Technol., 1–19 (2012)
2. Bühler, C.: Ambient intelligence in working environments. In: Stephanidis, C. (ed.) UAHCI 2009, Part II. LNCS, vol. 5615, pp. 143–149. Springer, Heidelberg (2009)
3. Corchado, J.M., Bajo, J., de Paz, Y., Tapia, D.I.: Intelligent environment for monitoring alzheimer patients, agent technology for health care. Decis. Support Syst. 44, 382–396 (2008)
4. Dourish, P., Bell, G.: Divining a digital future. Mess and mythology in ubiquitous computing. MIT Press, Cambridge (2011)

5. Nihan, C.E.: Intelligent working environments, handling of medical data and the ethics of human resources. In: Omatu, S., Paz Santana, J.F., González, S.R., Molina, J.M., Bernardos, A.M., Rodríguez, J.M.C. (eds.) Distributed Computing and Artificial Intelligence. AISC, vol. 151, pp. 429–436. Springer, Heidelberg (2012)
6. Ehrwein Nihan, C., Firoben, L., Gonin, F., Hitz, M., Weidmann, J.: Les technologies intelligentes un risque ou une opportunité pour la GRH?, HR-Vaud, Lausanne (2012)
7. Floridi, L. (ed.): The Cambridge handbook of information and computer ethics. Cambridge University Press, Cambridge (2010)
8. Habermas, J.: Erkenntnis und Interesse. Merkur 213, 1139–1153 (1965)
9. Kinder, K.E., Ball, L.J., Busby, J.S.: Ubiquitous computing, cultural logics and paternalism in industrial workplaces. Poiesis Prax 5(3-4), 265–290 (2008)
10. Kujala, S.: Effective user involvement in product development by improving the analysis of user needs. Behav. & Inf. Technol. 27(6), 457–473 (2012)
11. Patten, K., Passerini, K.: From personal area network to ubiquitous computing: preparing from a paradigm shift in the workplace. In: Proceedings of the IEEE Wireless Telecommunications Symposium 2005, pp. 225–233 (2005)
12. Ramos, C., Marreiros, G., Santos, R., Freitas, C.F.: Smart offices and intelligent decision rooms. In: Nakashima, H., Aghajan, H.K., Augusto, J.C. (eds.) Handbook of Ambient Intelligence and Smart Environments, pp. 851–880. Springer, Heidelberg (2010)
13. Röcker, C.: Acceptance of future workplace systems: how the social situation influences the usage intention of ambient intelligence technologies in work environments. In: Proceedings of the 9th International Conference on Work with Computer Systems, WWCS 2009, CD-ROM (2009)
14. Sousa, J.P.: Foundations of team computing. Enabling end users to assemble software for ubiquitous computing. In: Proceedings of the 2010 International Conference on Complex, Intelligent and Software Intensive Systems, CISIS 2010, pp. 9–16. IEEE Computer Society, Washington DC (2010)

Conversational Agents as Full-Pledged BDI Agents for Ambient Intelligence

Aida Mustapha[1], Mohd Sharifuddin Ahmad[2], and Azhana Ahmad[2]

[1] Faculty of Computer Science and Information Technology, Universiti Putra Malaysia,
43400 UPM Serdang, Selangor Darul Ehsan, Malaysia
aida@fsktm.upm.edu.my
[2] College of Information Technology, Universiti Tenaga Nasional, Jalan IKRAM-UNITEN,
43009 Kajang, Selangor Darul Ehsan, Malaysia
{sharif,azhana}@uniten.edu.my

Abstract. The need for conversational agents based on the Belief-Desire-Intention (BDI) architecture rise to support natural interfaces in many application areas of ambient intelligence, among which are the smart homes, health monitoring and assistance, care for the elderly, transportation, education, and tourism. The objective of this paper is to present an implementation of a conversational agent as a full-pledged BDI agent. This is achieved through mapping the BDI constructs in intelligent agents to the natural language processing components in a classic conversational agent, Eliza. Discussions focus on the process of reengineering the conversational agent technologies into BDI approach.

1 Introduction

The need for conversational agents based on the Belief-Desire-Intention (BDI) architecture rise to support natural interfaces for ambient technologies such as the autonomous car, smart homes, consumer electronics including the mobile phones, clothings and many others. Existing implementation of whether speech-based or text-based dialogue systems are primarily developed upon natural language processing technology such as natural language understanding, dialogue management, and natural language generation. It is often tied to a specific domain to manage the dialogue management and information database.

Rather than a rigid architecture, the driving key to future ambient application development should capitalize on the concept of agency, hence the autonomous capability. Autonomy is distinguished by the need to make decisions at any given time based on the context of current situation, hence autonomous systems are designed to make a rational evaluation of the choices available including the possible courses of action that could be taken. The level of autonomy ranges in capability from fully autonomous systems that are capable to make own decision without human control, semi-autonomous systems that requires certain level of approval from human authority or intelligent assistants that provide advice but takes action based on decisions made by human.

Being autonomous, intelligent agents work independently towards accomplishing explicit goals in making decision. In 1989, Pollock [11] introduces the terms cognitive and conative side of an intelligent agent in general. The cognitive structure contains declarative and procedural knowledge about the agent and its universe, and sets

A. van Berlo et al. (Eds.): *Ambient Intelligence – Software & Applications*, AISC 219, pp. 221–228.
DOI: 10.1007/978-3-319-00566-9_28 © Springer International Publishing Switzerland 2013

to achieve the built-in goals. The conative structure sets the preference or desire of the agent. His argument is that an agent is only interested in its own state in the universe, not the general state of the universe. Given this premise, an agent is likely to ascertain its own circumstances in making decision as influenced by its conative structure. From the perspective of intelligent agents, this means a conversational agent requires a self-reflecting thought mechanism or desire in its conative structure prior to communicating with its knowledge base by wiring the agent with (1) sensory input that represents its belief, and (2) rational deliberations that result in its intentions.

The conative structure of agents has been formalized as belief (B), desire (D), and intention (I) by [12]. Belief-Desire-Intention (BDI) is a mature and commonly adapted architecture based on the theory of human practical reasoning by Bratman [3]. Following the BDI architecture, agents are supplied with a set of plans (intentions) that describe means to achieving their goals (desires) under varying circumstances as triggered by events according to their internal informational state (beliefs) of the world. In the essence, a BDI agent adopts a plan as its intention in reaction to an event that could come from a change in environment or change in its belief. The algorithm for BDI interpreter is as follows, which to date has benefited from various extensions to support capabilities [9], look-ahead planning [13], commitment [4], and norms [8] to state a few.

The objective of this paper is to present an implementation of a conversational agent as a full-pledged BDI agent. This is achieved through mapping the BDI constructs in intelligent agents to the natural language processing components in conversational agent. The remaining content is organized as follows: After the slight insight in conversational agency, conation, and BDI formalism, we will present an existing conversational agent called Eliza [14] and the software Jack Intelligent Agents that is used as the vehicle to implement Eliza into a full-pledged BDI agent. Discussions that follow focus on the process of reengineering the conversational agent technologies into BDI approach. Finally, we will conclude the findings with some indication for future research.

2   Conversational Agents

Research has shown that the use of conversational agent results in improved understanding of the presented information [2]. Conversational agents exploit natural language technologies to engage users in text-based question-answering [10] or task-oriented dialogues for a broad range of applications [7]. It is also an intuitive interface for technologies and applications using ambient intelligence such as in care homes, workplaces, hospitals or schools. Three basic components in a conversational agent are (1) the interpreter to perform parsing on user utterance, (2) the dialogue manager to manage content and turns of the dialogue structure, and (3) the response generator to generate response utterance back to user. Different levels of analysis by the interpreter are essentially the natural language understanding (NLU) tasks and the response generator is part of natural language generation (NLG) process. Meanwhile, data stores are the domain-specific knowledge base with authoring facilities that should be straightforward enough to be maintained by non-technical personnel. Figure 1 shows components of conversational agent adapted from [7].

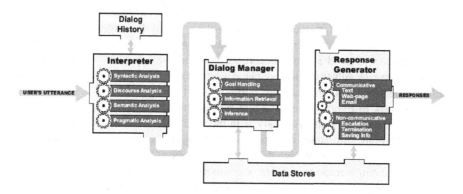

Fig. 1. Components of Conversational Agent [7]

This paper does not intend to model complex natural language processing techniques from tokenization to generation because the objective is to re-implement a conversational agent in a BDI architecture regardless of its natural language processing technologies. Hence, a conventional conversational agent called Eliza [14] is chosen due to its architectural simplicity. Eliza is perhaps the first computer program that is able to communicate with human user in natural language. In his article, Weizenbaum [14] described the logic of natural language processing capabilities in Eliza. The behavior of Eliza is controlled by a script that consists of input patterns extracted from user utterance and corresponding responses, which is also appended at the end of the article.

The core of natural language processing technology in Eliza lies only in text manipulation based on the concept of transformation rule associated with certain keywords. The transformation rule technique, as described in detail by [14], works to (1) decompose an input string into a selection of keywords, and (2) reassemble the decomposed string based on given assembly patterns. An example of a decomposition pattern with the corresponding assembly pattern is given below, where one keyword may have more than one decomposition pattern as well as reassembly pattern and (*) represents an indefinite number of unrecognized words:

Input string: (1) It seems that (2) you (3) hate (4) me.
Keyword: you
Decomposition pattern: * you * me *
Reassembly pattern: What makes you think I (3) you?
Output string: What makes you think I hate you?

The natural language processing steps are as follows. First, the entire user utterance is tokenized into whole words separated by spaces. This means no stemming or reducing inflected/derived words into its root are performed. Second, the list of keywords is constructed from the words that have been tokenized and is sorted according to the weights associated with the keywords in the script. In the example above, the first recognized keyword found from the list is the word "you". Third, for the given keyword, a decomposition pattern is selected, which has the word "you" and "me" with infinite number of unrecognized words in between. If no pattern is found, Eliza will repeat this process using the second keyword in queue. Fourth, for the selected decomposition pattern, a reassembly pattern is selected. In the script, there may be more than one reassembly pattern and Eliza will use the pattern according to its sequence. Finally, the resulting response utterance after substitutions take place is returned to the user with the impression that Eliza understood what the user was implying.

Referring back to Figure 1, text manipulation techniques in Eliza show that Eliza only performs a shallow syntactic analysis in the interpreter module. The transformation rule of decomposition and reassembly is working in the capacity of the dialogue manager that is supposed to manage the state of dialogue as well as the dialogue strategy. Finally, the entire response generator component is not applicable because Eliza response is based on substitutions in the reassembly pattern templates. The following section will present the mapping of BDI constructs to the component of conversational agent Eliza.

3   BDI Implementation

Programming Eliza in BDI is implemented using Jack Intelligent Agents (Jack) by Agent Oriented Software [6], [15], which is a Java-based agent-oriented development environment. While the term agent may describe a range of software components, the entire system in agent-oriented programming is modeled in the form of agents with autonomous reasoning entities capable of making pro-active decisions while reacting to events in their environment [1]. Jack is based on the Belief-Desire-Intention (BDI) paradigm that has been extensively studied in [12] and [16] and fully supports the design, implementation, and monitoring of BDI agents.

The BDI paradigm is an important feature of Jack using agent-oriented concepts such as Agents, Event, Plans, and Capabilities as well as Agent Knowledge Bases. The Agent class embodies the functionalities associated with Jack, the Event class is the originator of all activities within Jack, and the Plan class describes a sequence of actions that an agent can take when an event occurs. The core programming constructs in Jack is shown in Table 1. The constructs Agent, Event, and Plan in Jack are graphically represented by

Table 1. Core Jack Programming Constructs

Construct	Description
Agent	Agents are the individual computational entities within Jack.
Event	Events are the prime mover or the central motivating element in agents. Events are generated in response to external stimuli or as a result of internal process.
Plans	Plans are a set of procedures that define how agents should respond to events. When an event is generated, Jack will consider all plans applicable to the event and select the plan as its next intention. There may be more than one plan to achieve the same goal.
Beliefset	Beliefsets represent the agents belief as programmed in Jack. This is knowledge about self and the world it resides in.

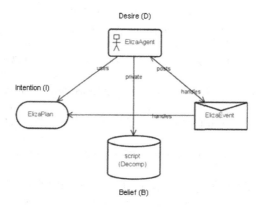

Fig. 2. Design Eliza_Ad in Jack Environment

icons connected with lines that indicate the relationships between the constructs, which
include the goal, contexts, reasoning steps, and actions by the agents.

As described, each Jack agent has (1) a set of beliefs about the world (its dataset),
(2) a set of events that it will respond to, (3) a set of goals that it may desire to achieve,
and (4) a set of plans that describe how it can handle the goals or events that may arise.
Figure 2 shows the design for Eliza as implemented in Jack, where (1) Eliza has a script
as its belief, (2) Eliza posts and handles event from user, (3) Eliza has desire to respond,
and (4) Eliza uses plan. As a conversational agent, the design shows that when Eliza is
instantiated, it will wait for input utterance from a user that it should return a response
to. When such input is obtained, it is considered as an event that needs to be resolved
hence Eliza will look through in its plan base on how to respond to the event.

Belief (B) represents the informative component or knowledge of the agent state
(self and the world it resides in) that is stored in a beliefset. A belief first come from
an environment, i.e., the agent believes that an utterance exists in its environment and it
needs to respond. Beliefs may also be implemented as inference rules that lead to new
beliefs. In Eliza, belief is implemented as a set of decomposition rules (Decomp) in the
script file. The script contains Decomp rules that extract the pattern in user utterance
and determine which reassembly pattern will be used to construct the response. The
decomposition and reassembly patterns represent its new belief as a consequence of the
Decomp rules (function).

The following shows the decomposition and reassembly patterns as extracted from
script for a keyword "you". A given keyword may have more than one decomposition
pattern and one decomposition pattern may have more than one reassembly patterns as
described by Weizenbaum [14].

Decomposition pattern: (1) you are (2)
Reassembly pattern:
    What makes you think I am (2)
    Do you sometimes wish you were (2)
    Perhaps you would like to be (2)

Desire (D) represents the motivational component of the agent state in the form of goals to be accomplished. A conversational agent desires a list (plan) of possible responses to the user, whereby the list represents the many ways the agent could take to achieve its goal. In Eliza, desire involves deliberating for the possible reassembly patterns from decomposing the user utterance until a keyword is extracted, matching the keyword through its belief coded in Decomp, adopting a reassembly pattern as its intention, and finally returning a response utterance back to user. However, note that one of the important aspects of the BDI reasoning model is that it models goal-directed behavior in agents, rather than plan-directed behavior [1]. This means a conversational agent will commit to return a response to user (its desire), regardless the methods chosen to achieve it.

The following is the parsing algorithm in Eliza adapted from [14] as explained in Conversational Agent. Line 1 to Line 16 is executed each time Eliza gets an event (input utterance) from user and its belief is updated. The entire algorithm concerns on deliberating for the best reassembly pattern.

```
01 perform input transformation
02 decompose input into array of words
03 repeat for each word
04    match with keyword in script
05    if keyword found
06       try each decomposition pattern in sequence
07       if match with keyword
08          execute reassembly pattern and return
09       else
10          repeat line 6
11    else
12       search keyword in memory
13       repeat line 5
14    else
15       return random response
16 end repeat
```

Finally, intention (I) represents the deliberative component of the agent state that have been committed based on plans (possible reassembly patterns). In other words, what makes Eliza choose the best response is its intention. Every time Eliza posts an event to indicate a user utterance is received, ElizaPlan is the first source that Eliza will consult to proceed in achieving its desire. Plans can be thought as a procedural manual or a collection of methods in conventional programming languages. However in Eliza, planning only involves interaction between the decomposition and reassembly patterns, dictionary of associative words or the backup of random answers. Because all are implemented in the single-point script file, hence Eliza only needs one plan. Due to this, the BDI mapping in Eliza may appear to be too simplistic. The following section will discuss the potential of a BDI-based conversational agent.

4   Discussion and Conclusion

Recall that the beliefset for Eliza Agent is the Decomp script, which is a straightforward implementation for pattern matching that sidesteps the entire response generation process as in conventional natural language generation component. However, this is not necessarily the de facto implementation of conversational agent. The response generator may be implemented as another agent that is responsible to generate response utterance using information in its beliefset. Meanwhile, the beliefset may be designed for a specific domain. For example, in a theater reservation system, the conversational agent must keep the details of theatrical performances such as title, date, show times, actor, director or other details. Similarly, in a Chemistry tutoring system, the beliefset may contain the details of Chemistry syllabus such as topic, equations, experiment or equipments. In a more advanced design, the beliefset may even be represented in the form of standard or domain-specific ontology under the responsibility of another ontology agent.

Unlike in Eliza, the separation of tasks for response generator agent and ontology-based beliefset promotes scalability and modularity in conversational agent development. The agent-oriented concepts could also be applied to other components in a dialogue system, such as for the interpreter. This means, more agents such as an ontology agent, thesauri agent or translator agent could be incrementally added to the design of conversational agent, while maintaining the architecture of belief, desire, and intentions. In this respect, the desire of a conversational agent should orchestrate execution of other supporting agents as strategized in its plan base, which is essentially the task of a dialogue manager. For instance, should the agent is unable to interpret user utterance, it may choose to ask the user to rephrase the utterance, consult the translator agent or connect to another ontology agent from different domain. Again, unlike Eliza, planning in conversational agent should not be limited to single reference point as in the script file.

Because a conversational agent is not able to carry out conversation about arbitrary subject, research on conversational agent or dialogue system is often being carried out in isolation. This is due to the architectural design that greatly varies from one system to another. Aside from the interpreter, dialogue manager and response generator, a dialogue system may consists input decoder such as for speech, handwriting or gestures and output renderer such a text-to-speech (TTS) engine, avatar or a talking head. From the academic perspective, implementation of dialogue system from component-based to agent-based is able to offer solution to the problem of subjective evaluation in conversational agents. This means for comparison purposes, researchers may choose performance of a particular agent, for example the recall and precision for the thesauri agent, while assuming other agents are working independently.

As for our future work, we believe that cognition and conation [11] are not all that there are to the mental attitudes, hence to the agent. Hilgard [5] first introduces the trilogy of mind that includes feeling and desire as well as knowledge, emotion and motivation apart from cognition and conation. In future, we will consider this third side of conversational agent, which is affection. We will proceed to implement the cognitive (thinking), conative (doing), and affective (feeling) self-regulating capabilities in conversational agent using BDI-based Jack Intelligent Agent. We hope this contribution will lead to realization of multi-modal systems with ambient intelligence that support speech, emotions, and body gestures in the near future.

Acknowledgement. This research is funded by Fundamental Research Grants Scheme by Ministry of Higher Education Malaysia in collaboration with the Centre of Agent Technologies at University Tenaga Nasional, Malaysia.

References

1. AOS (2008) JACK Intelligent Agents Agent Manual, http://www.aosgrp.com/documentation/jack/agent_manual.pdf (cited January 10, 2008)
2. Beun, R.-J., de Vos, E., Witteman, C.L.M.: Embodied Conversational Agents: Effects on Memory Performance and Anthropomorphisation. In: Rist, T., Aylett, R.S., Ballin, D., Rickel, J. (eds.) IVA 2003. LNCS (LNAI), vol. 2792, pp. 315–319. Springer, Heidelberg (2003)
3. Bratman, M.E.: Intentions, Plans, and Practical Reason. Harvard University Press, Cambridge (1989)
4. Gaertner, D., Noriega, P., Sierra, C.: Extending the BDI Architecture with Commitments. In: Proceedings of the 2006 Conference on Artificial Intelligence Research and Development, pp. 247–257 (2006) ISBN 1-58603-663-7
5. Hilgard, E.R.: The Trilogy of Mind: Cognition, Affection, and Conation. Journal for the History of the Behavioral Sciences 16, 107–117 (1980)
6. Howden, N., Ronnquist, R., Hodgson, A., Lucas, A.: Jack Intelligent Agents: Summary of an Agent Infrastructure. In: Proceedings of the 5th International Conference on Autonomous Agents, Montreal, Canada, May 28-June 1 (2001)
7. Lester, J., Branting, K., Mott, B.: Conversational Agents. In: Singh, M. (ed.) The Practical Handbook of Internet Computing. Chapman & Hall (2004) ISBN-10: 9781584883814
8. Meneguzzi, F., Luck, M.: Norm-based Behavior Medication in BDI Agents. In: Proceedings of the 8th International Conference on Autonomous Agents and Multiagent Systems, Budapest, Hungary, pp. 177–184 (2009) ISBN 978-0-9817381-6-1
9. Padgham, L., Lambrix, P.: Agent Capabilities: Extending BDI Theory. In: Proceedings of the 17th National Conference on Artificial Intelligence and 12th Conference on Innovative Applications of Artificial Intelligence, pp. 68–73 (2006) ISBN: 0-262-51112-6
10. Prager, J.: Open-Domain Question-Answering. Foundations and Trends in Information Retrieval 1(2), 91–231 (2006)
11. Pollock, J.L.: How to Build a Person: A Prolegomenon. MIT Press, Cambridge (1989)
12. Rao, S., Georgeff, M.: BDI Agents: From Theory to Practice. In: Proceedings of the 1st International Conference on Multi-Agent Systems, San Francisco, CA, pp. 312–319 (1995)
13. Silva, L., Sardina, S., Padgham, L.: First Principles Planning in BDI Systems. In: Proceedings of the 8th International Conference on Autonomous Agents and Multiagent Systems, Budapest, May 10-15, pp. 1105–1112 (2009)
14. Weizenbaum, J.: ELIZA-A Computer Program for the Study of Natural Language Communication between Man and Machine. Communications of the ACM, 36–45 (1966)
15. Winikoff, M.: Jack Intelligent Agents: An Industrial Strength Platform. In: Bordini, R., Dastani, M., Dix, J., Seghrouchni, A. (eds.) Multi-agent Programming: Languages, Platforms and Applications, pp. 175–193. Springer, New York (2005) ISBN 0387245685
16. Wooldridge, M.: Reasoning about Rational Agents. MIT Press (2000) ISBN-10: 0262232138

Supporting Workers and Quality Management in Sterilization Departments

Stefan Rüther[1], Thomas Hermann[2], Maik Mracek[1], Stefan Kopp[2], and Jochen Steil[1]

[1] Research Institute for Cognition and Robotics
Bielefeld University, Universittsstr. 25, 33615 Bielefeld Germany
{sruether,jsteil}@cor-lab.uni-bielefeld.de
[2] CITEC Center of Excellence Cognitive Interaction Technology
Bielefeld University, Universitätsstr. 21-23, 33615 Bielefeld Germany
{thermann,skopp}@techfak.uni-bielefeld.de

Abstract. Sterilization of medical instruments is a complex task because of legal prescriptions and many regulatory specifications. Complicated instruments must be correctly disassembled and loaded on a rack before they can be automatically processed in a cleaning and disinfection machine. This paper proposes an assistive system helping workers to avoid fatal errors during reprocessing. The system provides disassembly instructions and important information for the reprocessing of medical instruments to attract the worker's attention to critical issues. Continuous improvements and context-sensitive instructions are integrated by the use of business process models, a tabletop projection system with a gestural interface and radio-frequency identification (RFID) for instrument tracking. The user interface adapts to the criticality levels of instruments, which are continuously gathered and updated from the quality management of a sterilization department.

1 Introduction

Central Sterilization Supply Departments (CSSDs) are facilities in hospitals and laboratories where contaminated medical instruments are cleaned, disinfected and sterilized. Reprocessing of medical instruments is critical, since mistakes can lead to dangerous residues on instruments that can cause harmful (nosocomial) patient infections, as well as higher costs due to faster instruments wear or repeated reprocessing. Legislation in Germany forces the manufacturer of medical equipment to deliver instructions for cleaning, disinfection and sterilization. Quality management and process documentation are also forced by law [8,11]. Although instructions and workflow definitions exist, the worker within a CSSD is not always aware of them for several reasons: First, new or inexperienced workers are not familiar with the different instruments and hygienic restrictions. Second, a time limit for reprocessing is given for a set of instruments to be sterilized and ready for use. Thus, time for checking the disassembly of instruments and the machine loading is limited. Third, there are several thousands of different instruments in a hospital. For each instrument the worker has to assure the correct disassembly and loading of a cleaning and disinfection machine. With the high number of different, complex and changing instruments this is a difficult task. Fourth, instructions are not available in a consistent way, because every vendor uses a different format. An

A. van Berlo et al. (Eds.): *Ambient Intelligence – Software & Applications*, AISC 219, pp. 229–236.
DOI: 10.1007/978-3-319-00566-9_29   © Springer International Publishing Switzerland 2013

assistive system can therefore be highly useful in order to help the worker with the re-processing of critical instruments by drawing his or her attention to correct and critical handlings. Failures during reprocessing can potentially be avoided by supporting the worker with the right information at the right time in the right place. Alongside, process documentation and evaluation also benefit from a system that enables fast input of process parameters, e.g. issues with a given instrument. Since a CSSD is a delivery service within a hospital, the system ideally embeds into the hospital logistics software. Fig. 1 shows the challenging environment in the decontamination area of a CSSD.

Fig. 1. Example of a workplace in the decontamination area of a CSSD. The worker disassembles contaminated instruments.

Given these requirements and the hygiene restrictions in the challenging decontamination area of a CSSD (Fig. 1), the development of a computer based user assistance system is an ambitious task. To the best of our knowledge no such system has been described for today's CSSD.

We propose an approach on how to support workers in the decontamination area of a CSSD with an assistive system that utilizes a depth camera for hand tracking as well as a user interface realized as a table-top projection to enable human computer interaction in this challenging environment. The user interface provides context-sensitive information about the instrument currently reprocessed and allows the worker to input process-relevant information, which can further be utilized by an additional quality management process to e.g. generate failure reports. Workflows and data management are controlled by executable business process models. We thereby address both the challenging task to display appropriate context-sensitive information to support the worker in the concrete task and the approach to integrate the assistance system in the larger context of the hospital's quality management and data infrastructure.

Sect. 2 gives an overview of related work. Sect. 3 describes the requirements for assistive technology in a CSSD. Sect. 4 introduces our our integrated system architecture that deals with these requirements and briefly describes the current implementation and interaction design. A summary and outlook is given in Sect. 5.

2   Related Work

The preparation of medical instruments for disinfection falls in the broad class of manual tasks which in the literature are mostly discussed in the context of assembly and manufacturing. In particular, the project ACIPE focuses on intuitive and naturalistic interaction between a worker and an assistive system for human manual workplaces. It proposes mental, cognitive and process models for adaptively presenting instructions, according to working situations [2]. The "Attentive workbench" assists the worker in manual assembly with automatic parts delivery and projection of assembly instructions [9,10]. Ziola et al. [13] propose an augmented reality system for guidance with the assembly of LEGO Bricks. Additionally, the system detects constructed objects via computer vision and augments the environment with multimedia content. Zhang et al. [12] combine RFID, inertial sensors and a head mounted display to support the assembly of a 3D-puzzle and a computer mouse.

This literature describes assistive technology more or less applicable in the CSSD-domain, but is limited in relying on a fixed set of predefined data for worker guidance. Changes to this data require a separate interface and cannot be made within the interaction loop. The question on how an assembly system can acquire and maintain a valid set of instructions remains open. This is a critical question for the application of assistive systems in manual assembly and especially in hospitals, where thousands of different medical instruments exist and change over time.

3   Interaction Requirements and Hardware Setup

An assistive system for worker support in the decontamination area of a CSSD must meet special requirements. Acceptance and usability of the system by the worker is critical, because otherwise the worker will most likely ignore the system. This issue implies a number of requirements for placement of instructions and relevant information in the workers field of view as well as the robust and fast possibility for user input and instrument annotations. Interaction design is important, because context-aware information and context-aware control elements have to be present in the worker's field of view and thereby attracting the worker's attention to the critical tasks of reprocessing a complex instrument. However, in case of simple and uncritical instruments (e.g. clamps) the worker does not need assistance and therefore must not be annoyed. For the selection of interaction hardware, keyboards and touch screen are impossible in the decontamination area of a CSSD because the workers uses thick hygienic gloves, contaminated with residues from the operating room and cleansing material. Moreover, ergonomics and organization of the workplace are important because they directly influence the process efficiency and the worker's acceptance. Consequently, a touch-less interaction is the method of choice, which we implement by means of a gesture controlled projective display. Additionally, the time for reprocessing medical instruments should not be delayed by the use of the assistance system.

Our system setup is shown in Fig. 2(a). For our prototype, we assume that RFID-tags are mounted on each instrument, which allows tracking and automatic process documentation for single instruments. Process documentation is required by law, as

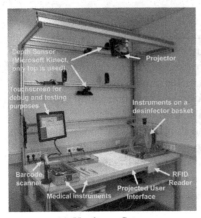

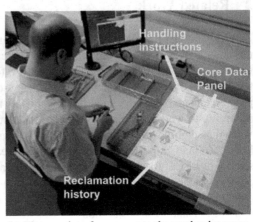

(a) Hardware Setup

(b) The user interface supports the worker by reprocessing instruments. All UI-elements are visible.

Fig. 2. System Prototype: The UI is projected on the workspace. A depth camera tracks the worker's hand and allows interaction with the UI. Additionally installed RFID and Barcode scanners are used for instrument tracking and process documentation.

already described in Sect. 1. Our prototype also includes also a barcode scanner for tracking sets of instruments, which is already widely used in practice.

The RFID-reader is mounted in the front of the workstation. A projector displays information directly on the workspace and a depth camera captures the workspace from above. This hardware setup provides a gesture controlled projective display. Fig. 2(b) shows the User Interface (UI) which is context-sensitive to the instrument's criticality level. Only a small number of UI-elements are used: A panel shows instrument core data while two other optional panels display instructions and special issues (e.g. reclamations) for an instrument. The visibility and configuration of these UI-elements are controlled by workflow models.

4   System Architecture and Prototype Interaction

Our system uses the Modell-View-Control pattern [3]. The business process models, a database with instruction and process information and a functional UI-model hold the necessary information about the overall process as well as the process logic and UI states. The UI displays this information and a controller executes the business process models and controls the application. In the following, we describe the basic workflow and the system architecture.

In the proposed system, the worker starts reprocessing of an instrument by scanning the instrument's RFID-tag which is a unique identifier of the instrument. Default instructions for reprocessing are previously assigned to the instrument's RFID-tag by a CSSD administrative with a desktop pc running a dedicated application ("AdministrationGUI"). The worker can annotate these initial instructions during reprocessing assisted by the UI "InstructionGUI" in order to report process-relevant data, e.g. reclamations on defect instruments.

The system guides the worker through the disassembly and rack loading as shown in Fig. 2(b). The framework "dSensingNI" [6] processes the image from the depth camera and extracts the worker's hand positions and gestures which it communicates via the TUIO protocol [5]. The system supports trigger actions analogue to a mouse click, i.e. a 'touch-less finger event' (defined by hand sufficiently close to table) allows initiating a software function connected to a button or widget element at that location. Direct touches with the contaminated surfaces in a CSSD can be avoided by specifying a desired distance between the hand and the surface.

The CSSD quality management is utilized to classify instruments' criticality: The operating room or a CSSD worker annotates an instrument if any problem such as a defect or wrong assembly occurs. Our system uses this reclamation history of a given instrument to influence the interaction behavior. A business process model ("CriticalityClassification") continuously classifies the criticality of the reclamation history: If a specific instrument had several issues in the past, the workflow model ensures that precise reprocessing instructions are displayed in the worker's field of view. Additionally, a confirmation dialog must be acknowledged, to assure that the worker is aware of the typical issues with the instrument. In contrast, no information is shown for instruments classified as non-critical, because we can assume that the workers are already familiar with the correct reprocessing, and therefore they are not disturbed by unnecessary information and dialogs. If there is no or not enough information available for a specific instrument, the system automatically asks a CSSD administrative to submit reprocessing instructions. The technical challenge is to integrate the assistance system into the overall quality management of the hospital. The CSSD processes can be modeled very well, since legislation and reprocessing guidelines exist. By utilizing business process models and the powerful "Activiti"-framework [7] for the definition of our UI-behavior, the application potentially integrates smoothly with other processes in the CSSD and the hospital, e.g. purchase logistics.

Data maintenance is done by BPMN 2.0 process models [1,4]. The "Activiti"-framework provides an execution engine which allows to model the CSSD process in detail and to bind software functionality to the different states of workflow. As Fig. 3 shows, we use BPMN 2.0 models to 1) acquire instruction datasets in a consistent representation[1], 2) to assess the criticality of these instructions for the worker, 3) control the user interface states regarding to the criticality level and 4) get reclamations and annotations on instruments from workers consistently. More in detail, the table top UI software architecture uses a Model-View-Control pattern. The model represents the UI-State and is controlled by the business process models. Since the Activiti-framework allows graphical edition of business processes, the behavior of the UI can be modeled graphically. Simplified, the UI-model for the workspace user interface has one property for an information object that is shown and one for the information density on the table.

---

[1] We defined a domain specific data object 'InfoStruct' that holds all information about an instrument such as core data, instructions, reclamation history, etc.. This data struct uses standardized data formats for multimedia instructions (e.g. JPEG, AVI, PDF ...) and forms the main data object for internal representation of instruments' data.

The information density conditions how much details are presented on the workspace and which user control elements like buttons are necessary. The lowest information density is a simple representation of a help button and basic information about the current instruments core data, like name and ID. The highest information density provides additional instructions for reprocessing as well as warnings in case something went wrong with the instrument. The instruction can be presented in video, picture and text form. With the use of business process models, very complex tasks can be modeled and the UI-state can be set for each subtask. For instance, a process model describes the subtask of an instrument's disassembly process and the UI-states and confirmation dialogs for each subtask.

In our prototype, scanning an instrument's RFID-tag starts the business process "TaskManagement" which first checks if there are previous tasks that have to be accomplished before the new instrument can be handled. In case of unfinished previous tasks, the process model either automatically marks the task as 'done' or it shows a special confirmation dialog. In this way, tasks are auto-accomplished, if the previous instrument was classified as uncritical or the confirmation dialog appears if a previous task on critical instrument has not been accomplished yet. This ensures the worker's attention on critical issues. After acknowledgement of the previous task is guaranteed, a workflow for setting the UI-model properties of the reprocessing view is executed. Based on the information available and the reclamation status of the given instrument, the UI is set to one of three modes. In the 'silent'-mode, nothing but core data and a help button is shown. The 'warning'-mode is used when the reclamation count is greater than one and a message is shown about the last issues with the instrument. The mode 'critical' is used when more than two mistakes happened during the last reprocessing. This mode additionally enforces the worker to explicitly acknowledge the correct treatment of the instrument. Symbolic weather-icons inform the worker about the criticality of the current instrument. During reprocessing, predefined annotations can be selected in order to add a reclamation on an instrument, e.g. if the worker wants to mark an instrument as wrong assembly. Adding a reclamation starts a business process, which calculates and updates the criticality of the instrument.

While developing and testing our prototype, we have conducted pre-studies with students from computer science. With no knowledge about the system and the reprocessing of medical instruments, they were able to correctly prepare medical instruments for automatic cleaning and disinfection. They did not achieve this without the system. Instruments were correctly disassembled and correctly loaded on a cleaning rack, reclamations could easily be added to a specific instrument. Qualitatively compared to reprocessing without assistance provided by the system, they did fewer errors and needed the same amount of time.

These very preliminary but encouraging experiences show the potentially high effectiveness for supporting workers. A quantitative study to measure time, error rates, acceptance and usability of the system compared to the situation found in today's CSSDs is currently under the way.

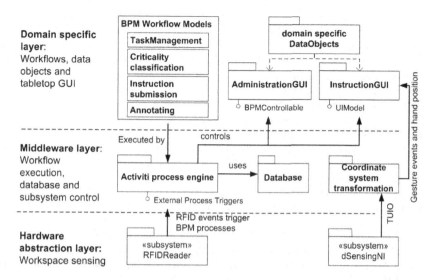

Fig. 3. System architecture. Workflow models are executed by the process engine and control the persistent data storage as well as the user interface for instruction submission and instruction view. The domain specific layer contains specific data structure and subsystems for the CSSD domain. The 'AdministrationGUI' is currently used for the input and editing of instructions.

5   Conclusion

Our proposed approach for assisting CSSD workers combines BPMN 2.0 workflow modeling together with gesture-based interaction and tabletop projection in order to assist the worker with the complex cleaning, disinfection and sterilization of medical instruments. The interactive display is controlled by BPM processes and therefore directly integrates in a larger scale application that controls the overall processes in a CSSD domain, e.g. quality management. A prototype of the proposed system shows high potential for effective worker guidance. The system allows to aggregate information about preparation issues and equipment problems that can be accessed and shared between several workplaces. Future work will extend the prototype's functionality and enhance the integration in larger scale CSSD software and future studies will evaluate the usability and failure prevention capability of the prototype.

References

1. Axway, Machines, I.B., International, M.E.G.A., Oracle, Sap, A.G.: Unisys: Business Process Model and Notation (BPMN) Specification 2.0 (2009)
2. Bannat, et al.: Towards Optimal Worker Assistance: A Framework for Adaptive Selection and Presentation of Assembly Instructions (2008)
3. Buschmann, F., Meunier, R., Rohnert, H., Sommerlad, P., Stal, M., Stal, M.: Pattern-Oriented Software Architecture vol.1: A System of Patterns, 1st edn., vol. 1. Wiley (1996)
4. Chinosi, M., Trombetta, A.: BPMN: An introduction to the standard. Computer Standards & Interfaces 34(1), 124–134 (2012), doi:10.1016/j.csi.2011.06.002

5. Kaltenbrunner, M., Bovermann, T., Bencina, R., Costanza, E.: TUIO - A Protocol for Table Based Tangible User Interfaces. In: Proceedings of the 6th International Workshop on Gesture in Human-Computer Interaction and Simulation (GW 2005), Vannes, France (2005)

6. Klompmaker, F., Nebe, K., Fast, A.: dSensingNI: a framework for advanced tangible interaction using a depth camera. In: Proceedings of the Sixth International Conference on Tangible, Embedded and Embodied Interaction, TEI 2012, pp. 217–224. ACM, New York (2012), doi:10.1145/2148131.2148179

7. Rademakers, T.: Activiti in Action: Executable business processes in BPMN 2.0, 1st edn. Manning Publications, Shelter (2012)

8. RKI, BfArM: Anforderungen an die Hygiene bei der Aufbereitung von Medizinprodukten. Bundesgesundheitsblatt - Gesundheitsforschung - Gesundheitsschutz 44, 1115–1126 (2001), doi:10.1007/s00103-001-0279-x

9. Stoessel, C., Wiesbeck, M., Stork, S., Zaeh, M., Schuboe, A.: Towards Optimal Worker Assistance: Investigating Cognitive Processes in Manual Assembly. In: Mitsuishi, M., Ueda, K., Kimura, F. (eds.) Manufacturing Systems and Technologies for the New Frontier, pp. 245–250. Springer, London (2008), doi:10.1007/978-1-84800-267-8_50

10. Sugi, M., et al.: Quantitative Evaluation of Automatic Parts Delivery in "Attentive Workbench" Supporting Workers in Cell Production. Journal of Robotics and Mechatronics 21(1), 135–145 (2009)

11. Tabori, E.: Durchblick bei der Hygiene. Arthroskopie 21, 66–73 (2008)

12. Zhang, J., Ong, S., Nee, A.: RFID-assisted assembly guidance system in an augmented reality environment. International Journal of Production Research 49(13), 3919–3938 (2011), doi:10.1080/00207543.2010.492802

13. Ziola, et al.: Examining interaction with general-purpose object recognition in LEGO OASIS. In: 2011 IEEE Symposium on Visual Languages and Human-Centric Computing (VL/HCC), pp. 65–68 (2011), doi:10.1109/VLHCC.2011.6070380

An Intelligent System to Setup Meetings, Capture, Organize and Record Information in Smart Offices

Joaquim Teixeira, Carlos Lima, Lino Figueiredo,
Goreti Marreiros, and Ricardo Costa

GECAD – Knowledge Engineering and Decision-Support Research Center
Institute of Engineering – Polytechnic of Porto (ISEP/IPP)
Rua Dr. António Bernardino de Almeida, 431
4200-072 Porto
{jra,1091112,lbf,mgt,ricar}@isep.ipp.pt

Abstract. An ambient intelligent environment aims to have a pervasive computing, being a new paradigm that supports the projects of the next generation of intelligent systems and introducing new means of communication between man and machine. This paper reflects our efforts to develop an integrated control system for our Laboratory of Ambient Intelligent Decisions (LAID) in GECAD. The main purpose was to develop an application that is able to control three IP cameras Sony SNC-RZ25, together with the ability to monitor, control and view lab computers, thus adding a component of artificial intelligence in preparing and tracking a meeting in a smart environment. Therefore, this intelligent system is now closer to make decisions and interact with users without human intervention.

1   Introduction

AmI stands that in a near future, users are provided with services that support their activities in everyday life. AmI scenarios, like a smart meeting room, recognize the presence of individuals and seamlessly react to them. To accomplish this, some AmI systems are built with embedded sensors in the environment that acquire and exploit data with the purpose of generating automatic and semi-automatic responses to events. There are different sensor and network technologies: RFID, Wi-Fi, 4G cell networks, etc. However, visual sensors are receiving more attention in the last years, due to the ability of gathering a large amount of interesting data [1].

Videoconference or video surveillance is increasingly being used in events such as meetings. With the potential that the Internet offers us, using IP cameras, it is possible to visualize these same images anywhere in the world. Nowadays meetings are important events in our daily lives, where we can distribute information and share knowledge. Traditionally, in an old fashion meeting, people frequently forget important information presented, whether because they could not attend the meeting or simply because there's just too much data, information and knowledge to retain. Most of us take notes and review them later, a process that can be inefficient, incomplete and inconsistent. The smart meeting systems are designed to support a new paradigm,

A. van Berlo et al. (Eds.): *Ambient Intelligence – Software & Applications*, AISC 219, pp. 237–244.
DOI: 10.1007/978-3-319-00566-9_30   © Springer International Publishing Switzerland 2013

automatically recording meetings on digital media, to perform post tasks like archive, analyze and summarize a meeting [2, 3, 4]. In this article, we will start by exposing smart offices followed by the presentation of LAID, our test bed environment for this proposal. We will then present a new approach for video/other context data capture, preparing and tracking a meeting in a smart environment. A section with tests and results will be presented. This article ends with the conclusions and future work.

2   Smart Offices

To develop an intelligent environment, it is vital that the whole architecture efficiently uses the physical components such as sensors, software and intelligent electronic devices. Through these components, and using the information received by these sensors and software, it's possible to analyze the environment and launch actions in order to change its current state. For this, networks of sensors / actuators need to be robust and organized in order to create a platform for ubiquitous/pervasive computing. Streitz, et al., in 2005, developed a research where they were looking for an architecture that would support collaboration, informal communication and social awareness. The planning to this architecture was made with the purpose of giving components intelligence and the ability to interact with the user in a simple and intuitive way. These researchers have introduced the distinction between two types of intelligent components [5]. Systems Oriented: these create an environment with intelligent individual components and make it self-oriented performing actions based on the information gathered. Thus, the system leaves humans out of the decision process; People Oriented: the main objective is to ensure that "smart spaces make people smarter", so that the user has control of the entire system and can make the decisions, being the primary responsible for these same actions.

In an expedited basis, we can say that Smart Offices aim to support users in performing their daily tasks, making changes in the environment, thus allowing an individualization of the desktop for each user. The work described on this paper was developed based on this kind of environments and with an architecture "oriented to people", which, as stated before, leaves a great share of the responsibility to the user [10]. Figure 1 depicts a generic architecture for a smart meeting system where we can see three layers: Meeting Capture, Meeting Recognition and Semantic Processing, each and every one of them with some basic modules. The Meeting Capture layer is responsible for acquiring data from the environment, focusing on three major types of data: audio, video and other context data. The mid-layer, Meeting Recognition, is responsible to analyze the content captured by the Meeting Capture layer. It's where it resides services like speech recognition, person identification or activity recognition, giving some meaning to the captured content and at the same time providing the essential support to the upper layer, Semantic Processing. Semantic Processing handles high level manipulations such as Meeting Annotation, Meeting Indexing or Meeting Browser. The content of this article focuses on the two lowest layers, Meeting Capture and Meeting Recognition. We propose a solution based on a real scenario to deal with captured data using Video and Other Context Data, unifying these two sources of data in our lab at the Meeting recognition layer.

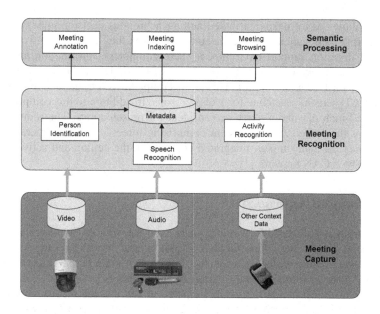

Fig. 1. Generic Architecture for a Smart meeting System (adapted from [6])

3   LAID – Laboratory of Ambient Intelligence for Decision Support

The tests performed to evaluate our proposal were done in the Laboratory of Ambient Intelligence for Decision Support (LAID) [7].

Fig. 2. LAID

LAID allows us to make distributed and asynchronous decision-making meetings, so the participants can attend the meeting wherever they are. The software in this intelligent decision support environment share computers, information and services throughout the network [8, 9]. Figure 2 shows GECAD's Laboratory of Ambient Intelligence for Decision Support. This lab is equipped with 6 interactive touch sensitive 26" LCD screens, each one for 1 to 3 persons. In addition to this equipment, there is also a 62-inch plasma, a touch screen ideal for presentations. The room is further equipped with a video / audio system comprising three CCTV IP cameras, multiple microphones, an audio mixer, integrated speakers in the ceiling and active terminals controlled by a CAN network [1].

4   Proposed Approach

This system was developed for a real scenario, the Knowledge Engineering and Decision Support Research Center (GECAD) LAID smart meeting room. As shown in figure 3, the system architecture consists of three new main parts: *Laid CamControl*, *iTALC* and a server for maintaining data from the meetings.

Laid CamControl is a new module prepared to capture and record the meeting as the time goes by, interacting with *iTALC*, an open source software for intelligent learning and teaching [11]. *iTALC* is able to give a lecturer permission to create classrooms, with computers and users. Once those computers from LAID and users are added, definitions can be saved and reused later. A person who is supposed to lead a meeting may have different needs than another one. That person can create a scenario where only 4 specific computers in LAID room are used, while another one might need all 6 computers. Therefore *iTALC* offers a lot of possibilities, such as see what's going on in computer-labs by using overview mode and make snapshots; remote control to all computers to support and help other people; show a demo (either in full screen or in a window) - the screen of who needs to expose something is shown on all other computers in real-time, and also can send text messages from the master computer to the others. Another very important feature: *iTALC*'s network-technology is not restricted to a subnet and therefore, people from outside LAID room at home, at work, on any part of the world can join the meeting via VPN-connections.

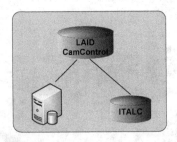

Fig. 3. Modified Partial LAID Architecture

This software is very important not to track a meeting but to support it. In figure 3, a connection with *Laid CamControl* can be seen. The interaction created between the two modules it's what gave us the ability to track a meeting and support it in an intelligent way, recording everything that happens, but essentially, recording what matters. If someone is speaking and its screen is transmitting data to the other ones, we need to focus our attention on that person and at the same time on what he or she is showing. *Laid CamControl* steps up at this time, being able to add, remove and control our LAID IP cameras. In the next figure, figure 4, *Laid CamControl*'s interface is showed. The primary camera is always the one transmitting at the center of the interface. The other ones can be added and stay at the left, with a smaller refresh interval, waiting for a switch to be made, by the user or automatically, turning, at the time of the switch, into the primary camera.

The Command Panel is located on the right side of the main window, as shown in figure 4. After binding to at least one camera, the user can control the orientation through the navigation buttons (enabled by default) or the bars angles (by selecting the slidebar control), both located on the Command Panel.

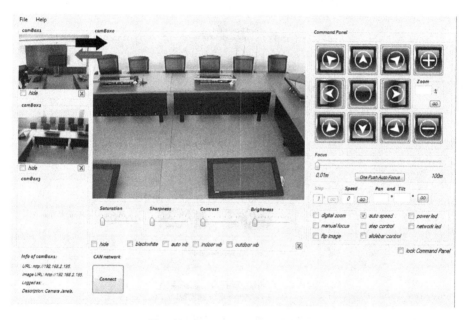

Fig. 4. *LAID CamControl* Interface

The application also has a scenario configuration functionality, allowing the user to choose between setting a new scenario or load a previously defined scenario. This feature has major advantages. One is that the user can set the location of the various terminals distributed by the conference room and several pre configured camera positions. This allows uploading one of several possible scenarios, composed of six, four or even two terminals, along with the positioning of all three cameras. The scenarios may still have the same number of terminals; however the layout of the tables can be changed for some reason if GECAD wishes. That information can be stored and be used at any time of the meeting, whether to initiate it, or to make quick, automatic and semi-automatic decisions to support the whole process of capture and record the meeting. Figure 5 depicts the overall system operation, since the meeting starts, explaining the flow when participants' interactions appear.

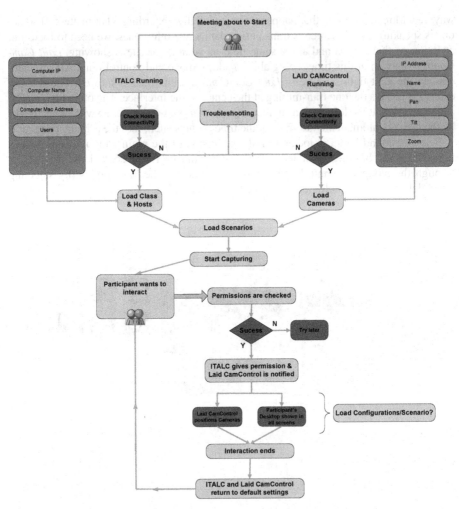

Fig. 5. Overall System Operation

When a meeting is about to start, *iTALC* and *Laid CamControl* must be running on the main Desktop. *iTALC* checks the connectivity with all the hosts existing in LAID and loads a user class and respective hosts. Each host has an IP, name and MAC Address. This communication between the main desktop and the clients is secure, using credentials, public and private keys to exchange authentication data. At the same time, *Laid CamControl* is responsible to check if all the cameras are reachable and load them. These cameras have an IP address, a name, as well a default value for PAN, TILT and ZOOM. The second main step is to load a scenario, accordingly to the type of user that is responsible for the meeting. Customized scenarios can be stored and loaded, containing witch terminals will be used and witch one is the primary (a meeting might not necessarily start by someone presenting documents or media in the 62'' plasma); initial values for pan, tilt and zoom of all cameras are also loaded. The meeting starts to be captured with a central camera responsible to collect video signals from what's being showed at the 62" plasma and from the area around this equipment, where usually a person might be. The other two lateral cameras are prepared to follow the other participants when they interact with the systems and present relevant information to the meeting.

This leads us to a final and third phase presented in figure 5: participants' interactions. When someone wants to interact, by speaking and specially present documents or any other kind of media to the other participants, *iTALC* must grant or deny permission. If it is granted, *Laid CamControl* is notified and cameras are repositioned as well as all screens get to see what that person has to show. This can be optimized by doing something very similar when loading scenarios at the beginning. By example, if terminal 2 was granted permission, values for pan, tilt and zoom of all cameras might be loaded automatically, without human intervention, once there was a prior configuration and the system may react by itself to the change.

5   Implementation and Testing

All implementation was done with C# and C++ programming languages, plus additional CGI scripts in order to process HTTP requests to the cameras. Several tests were performed in order to assess the system's capacity in dealing with IP cameras and communication between the two modules. Through an implementation layer perspective, boundaries had to be defined when a camera needs to move by itself. Values of tilt, pan and zoom must be restricted in a first approach and then controlled when communicating with the remote server embedded in each camera, so that we can guarantee the correct positioning from a whole scenario to a simple movement.

When dealing with IP cameras and software that is able to communicate among an IP network, some cautions need to be taken care. Besides the inserted common controls that improve the system's usability, such as trying to add cameras with wrong IPs or IPs that do not belong to the network itself, system performance was a continuous concern throughout the development of this work. Communication is fundamental between the two modules presented, but also between *Laid CamControl* and all three IP cameras. Also, it was very important no to overload the remote server of each camera with unnecessary processing and requests. Modifications were made throughout the development of the work core: at the beginning delays were frequent and most of them were treated. It is critical that all the equipments that need to send packets to the network stay in the same network segment, preferably at the same switch in the network physical level. This way, fewer hops are necessary to a packet to reach its destination, and in cases such as our work, involving technologies similar to remote control like VNC (iTALC) and streaming (IP cameras), this is crucial.

6   Conclusions and Future Work

This paper described a solution that combines a new software module capable of control IP cameras, with the ability of interacting with an intelligent learning and teaching application, for a real scenario. Special focus was given to the ambient used as test bed, the Smart Office named LAID. After detailing our approach we have shown how our proposal is able to: capture and record video focused on the speaker and at the same time, on what all the meeting participants can see in real time. It is possible to monitor, control and view LAID lab computers during a meeting, adding an artificial intelligence component, that turns the system presented able to execute actions without human intervention. All the work done and presented at this article improved the capabilities of the intelligent systems already existent in LAID.

Future work can be oriented towards some limitations that weren't vital for this project: implementation of the controls for brightness, contrast and other various settings that can improve image quality. Facial/Voice recognition would improve the knowledge base, offering new possibilities when the time for analyzing the captured data comes, after a meeting is finished or even in real time. Emotional aspects can also be considered in the future, to support the decision making process.

Acknowledgments. This work is supported by FEDER Funds through the "Programa Operacional Factores de Competitividade - COMPETE" program and by National Funds through FCT "Fundação para a Ciência e a Tecnologia" under the project: FCOMP-01-0124-FEDER-PEST-OE/EEI/UI0760/2011.

References

1. Freitas, C., Meireles, A., Figueiredo, L., Barroso, J., Silva, A., Ramos, C.: Context aware middleware in ambient intelligent environments. Accepted for publication in Int. J. Computational Science and Engineering
2. Kynsilehto, M., Olsson, T.: Intelligent ambient technology: friend or foe?. Presented at the Proceedings of the 15th International Academic MindTrek Conference: Envisioning Future Media Environments, Tampere, Finland (2011)
3. Kim, H., Wolf, M.: Distributed tracking in a large-scale network of smart cameras. Presented at the Proceedings of the Fourth ACM/IEEE International Conference on Distributed Smart Cameras, Atlanta, Georgia (2010)
4. Gómez-Romero, J., Serrano, M., Patricio, M., García, J., Molina, J.: Context-based scene recognition from visual data in smart homes: an Information Fusion approach. Personal Ubiquitous Comput. 16 (2012)
5. Streitz, N., Rocker, C., Prante, T., van Alphen, D., Stenzel, R.: Designing smart artifacts for smart environments. Computer 38 (2005)
6. Yu, Z., Nakamura, Y.: Smart meeting systems: A survey of state-of-the-art and open issues. ACM Comput. Surv. 42(2) (2010)
7. Marreiros, G., Santos, R., Freitas, C., Ramos, C., Neves, J., Bulas-Cruz, J.: Laid – a smart decision room with ambient intelligence for group decision making and argumentation support considering emotional aspects. International Journal of Smart Home 2(2) (2008)
8. Marreiros, G., Santos, R., Ramos, C., Neves, J.: Context-aware emotion-based model for group decision making. Intelligent Systems 25(2) (2010)
9. Laranjeira, J., Freitas, C., Marreiros, G., Ramos, C.: A Digital Secretary for Smart Offices Setup Up. In: ISAmI 2011 (2011)
10. Sadri, F.: Ambient intelligence: A survey. ACM Comput. Surv. 43(4) (2011)
11. "iTALC", http://italc.sourceforge.net/home.php

Author Index